Windows to the World

Written and Illustrated
by
Nancy Everix

Cover by Vanessa Filkins

Copyright © Good Apple, Inc., 1984

ISBN No. 0-86653-173-4

Printing No. 987654

GOOD APPLE, INC.
BOX 299
CARTHAGE, IL 62321-0299

Dedicated with love to DeWitt and Mark for their support and encouragement.

TABLE OF CONTENTS

Introducing a...

World of Understanding

BRITISH COLUMBIA MEXICO GERMANY EGYPT

AUSTRALIA EGYPT BRITISH COLUMBIA GERMANY

Mexico

AUSTRALIA

Egypt

BRAZIL

China

BRITISH COLUMBIA CANADA

GERMANY

TIPS FOR THE TOUR GUIDE

Whether you are using all areas and plan a year's work or only a few days, some suggestions follow that may help your "tourists" enjoy their trip and keep their "luggage" together.

Use a grocery bag; glue construction paper on two sides. Decorate with student-designed airline stickers. Add a luggage tag made from a tile sample and cord. Use this carry-on bag for the tourists to keep their papers, artwork, and other information.

For a sturdier container (that may be used for more than one trip), cover a dress or shirt box with construction or self-adhesive paper. Add a cardboard handle and luggage tag. Again decorate with travel brochures, pictures, etc. Keep the student's passport, airline ticket, etc., inside.

Remember, your local travel agency and especially the travel agents may be your best resources for this entire unit. If you do use the services of an agency, please consider some kind of thank-you. Often we expect the agents to provide more than their job requires. Also, try not to give assignments which require a large number of students to invade the travel agency. You'll probably pay a double price for your next trip accommodations if you do!

There are addresses of foreign tourist agencies listed on the Free Materials page. Be sure to order these materials ahead of time, or better yet, order for the entire year. Then you and your class will not be disappointed!

LOCAL RESOURCE PEOPLE

Contact Person	Country	Address	Phone	Notes

Add to this list as you discover willing resource people. Keep it handy year round. Be sure to get permission from your resources before you give their names to a colleague.

PASSPORT APPLICATION

I, _____ ,
  (FIRST NAME)   (MIDDLE NAME)   (LAST NAME)

A CITIZEN OF THE UNITED STATES, DO HEREBY APPLY TO THE DEPARTMENT OF STATE FOR A PASSPORT.

☐ MALE ☐ FEMALE	BIRTHPLACE _____	BIRTH DATE MONTH | DAY | YEAR
HEIGHT ___ FT. ___ IN.	**COLOR OF HAIR** _____	**COLOR OF EYES** _____ / **SOCIAL SECURITY NO.** _____
STREET ADDRESS	CITY STATE	ZIP CODE

I SOLEMNLY SWEAR THAT THE INFORMATION GIVEN ABOVE IS TRUE AND THE PICTURE IS A LIKENESS OF ME.

_____   DATE _____ 19___
          SIGNED

                                         _____
                                         PASSPORT AGENT (TEACHER)

TRAVEL PLANS

PURPOSE OF YOUR TRIP	MEANS OF TRANSPORTATION SHIP ☐ AIR ☐ OTHER ☐
LENGTH OF STAY	HAVE YOU TRAVELED ABROAD PREVIOUSLY? YES ☐ NO ☐
DO YOU EXPECT TO TAKE ANOTHER TRIP? YES ☐ NO ☐	COUNTRIES TO BE VISITED:

PICTURE OF APPLICANT

PASSPORT

HOME SWEET HOME

Cut apart, assemble and fill in the needed information.

PICTURE

NAME _____

STREET _____

CITY, STATE _____

Height _____ Weight _____ Eye Color _____

Hair Color _____ Age _____

Agent + Tour Guide _____

DATE _____

COUNTRY _____

DATE ENTERED _____

DATE DEPARTED _____

GRADE ☐

Cut apart; assemble inside passport cover. Use one quarter page for each country studied. Glue the flag of that country onto the passport page.

COUNTRY _____

DATE ENTERED _____

DATE DEPARTED _____

GRADE ☐

COUNTRY _____

DATE ENTERED _____

DATE DEPARTED _____

GRADE ☐

VISAS

COUNTRY _____

DATE ENTERED _____

DATE DEPARTED _____

GRADE ☐

COUNTRY _____

DATE ENTERED _____

DATE DEPARTED _____

GRADE ☐

OUR WORLD FLIGHT PLAN

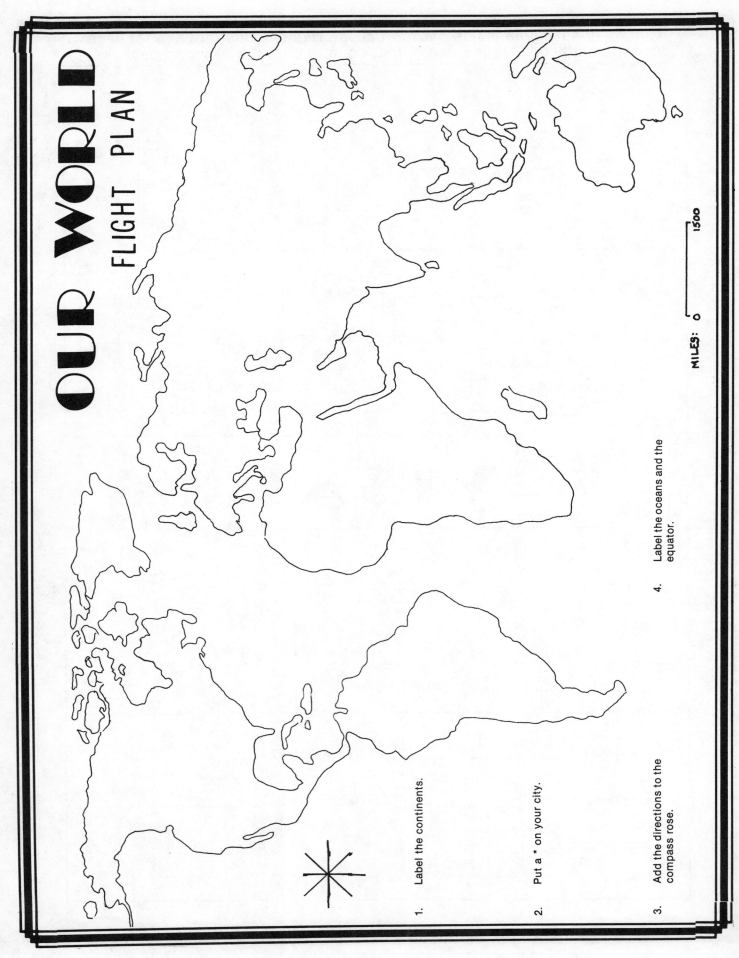

1. Label the continents.

2. Put a * on your city.

3. Add the directions to the compass rose.

4. Label the oceans and the equator.

MILES: 0 ____ 1500

AIRPLANE TICKET

Measure the distance to the country of your destination. Use the map of the world. Don't forget to use the scale of miles at the bottom of the map.

    How many inches is the distance?_____

    How many miles is this distance?_____

Determine the cost of your ticket, using the information above.
It costs 50¢ to travel 1 mile.

    *How much does a one-way ticket cost?_____

    *How much does a round-trip ticket cost?_____

*You may need a calculator to determine this cost.

Adventure Airlines

NAME _____   ROW _____

                                                   SEAT _____

DESTINATION _____

                                         MEAL YES ☐ NO ☐

DEPART TIME _____

ARRIVE TIME _____

                                         TOTAL COST _____

DISTANCE _____

FLAGS

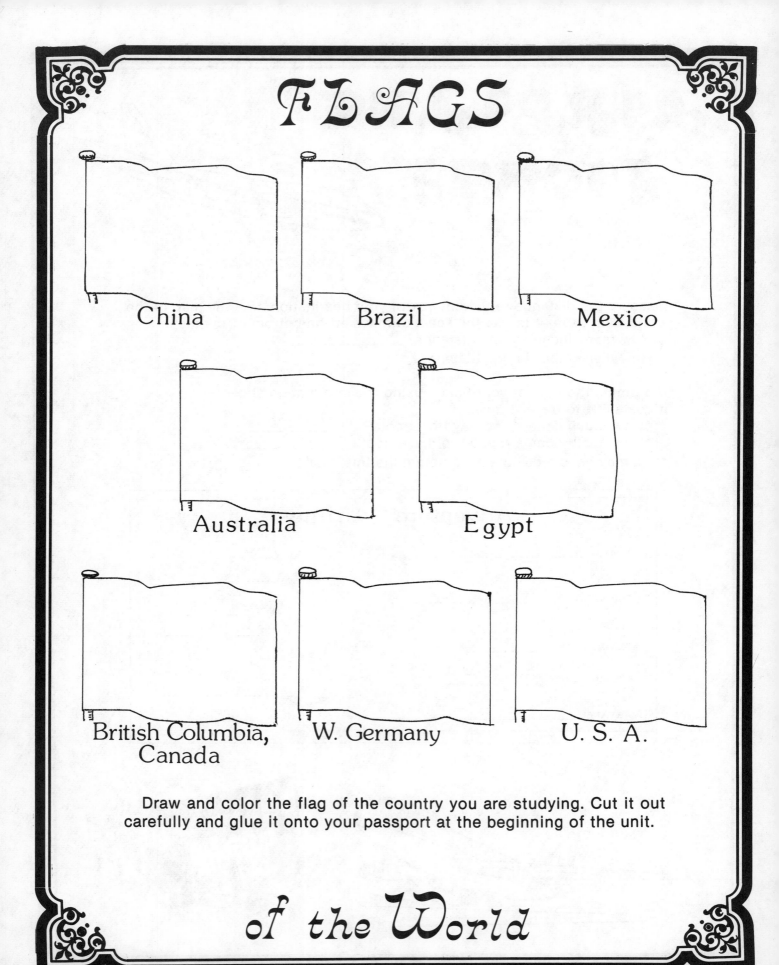

China

Brazil

Mexico

Australia

Egypt

British Columbia, Canada

W. Germany

U. S. A.

Draw and color the flag of the country you are studying. Cut it out carefully and glue it onto your passport at the beginning of the unit.

of the World

NATIONAL SEALS

Design a seal for each country you study. Include people, places, and things that are important to the specific country.

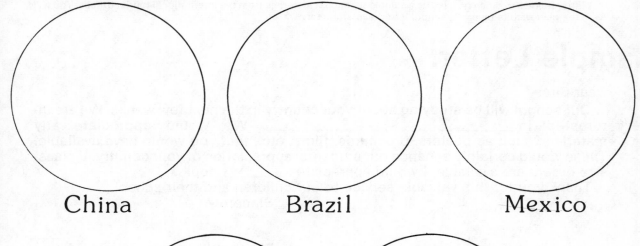

China

Brazil

Mexico

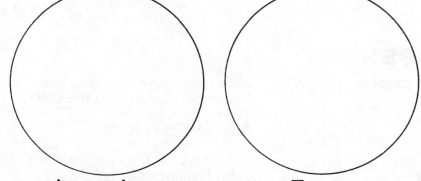

Australia

Egypt

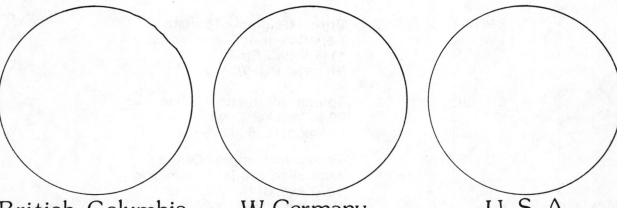

British Columbia,
Canada

W. Germany

U. S. A.

Cut out the seal and glue it onto the passport at the end of each unit.

of the World

FREE MATERIALS

Often you can receive class-size amounts of free materials and beautiful posters. Films and other visuals are often available for a nominal fee.

A letter follows that can be used as a sample to request materials from the international tourist offices. You will have the best results by using school letterhead stationery.

Sample Letter:

Dear Sirs:

Our school will be studying about your country in the next few weeks. We are interested in _____. We would appreciate any materials, such as posters, brochures, films, etc., that you would have available. These would certainly enhance our students' appreciation of your country. If class-size orders are available, I would appreciate _____ copies.

Thank you for this valuable service to our children and their school.

Sincerely,

Teacher
Grade Level

Addresses:

CHINA: Information and Publicity Dept.
Permanent Mission to the U.N.
155 W. 66th St.
New York, NY 10023

MEXICO: Mexican Government Tourism Office
John Hancock Center
875 N. Michigan Ave., Suite 3612
Chicago, IL 60611

BRITISH COLUMBIA: British Columbia Tourism
Department of Travel Industry
1117 Wharf Street
Victoria, BC V8W2Z2

BRAZIL: Tourist Information Office
20 N. Wacker Drive
Chicago, IL 60606

AUSTRALIA: Tourist Information Office
Australian Tourist Commission
1270 Ave. of the Americas
New York, NY 10020

GERMANY: German Tourist Office
104 S. Michigan Ave.
Chicago, IL 60603

EGYPT: Tourist Information
630 Fifth Ave.
New York, NY 10020

DESIGN YOUR OWN

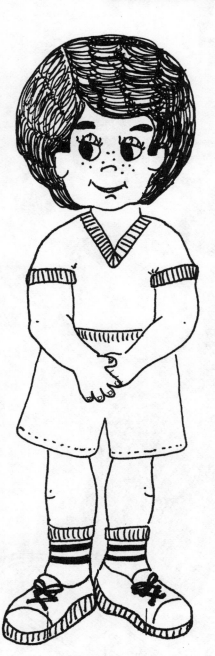

DOLL PATTERN: Duplicate these patterns on construction paper. Cut out. Glue 4-5 more pieces of construction paper behind each doll for thickness and support. Have children color these doll patterns. Cut slits with a razor blade in the feet for a stand.

ETHNIC COSTUME

TO MAKE THE STAND: Cut a rectangular piece of cardboard (1″ x 6″). Insert ends into slots to form a circular stand for the feet of the dolls. Add the ethnic costumes for the country that you are studying.

CHINA CHINA CHINA CHINA CHINA

CHINA

CHINA CHINA CHINA CHINA CHINA

TIPS
FOR THE
TOUR GUIDE

Klass Kickoff:

Explain to the students that you will be taking several interesting trips this year. However, you will not be traveling by school bus. The first trip will be by airplane to China. Have the students use the passport information and plane ticket to map the route on a world map. Have them determine how long the flight will take and what the climate is like. This will determine the clothing to pack. Then arrange the classroom to portray a ticket booth at the airport and the seating inside your 747 airplane. Use crepe paper streamers to connect the seats. Take volunteers to fill the jobs of ticket seller, stewardess, steward, pilot, copilot, navigator, ground crew, luggage handlers, and passengers. Now, reenact the trip: getting everyone on board, fastening seat belts, takeoff, pilot's dialogue of sights on the flight, navigator's calling off countries from a map, serving dinner, landing, collecting luggage, etc. Happy landing!

Tips:

China's official name is the People's Republic of China. This country has an area of 3,706,000 square miles and a population of approximately 1,003,555,000 people who speak the Mandarin Chinese and Cantonese languages. The economy is based on the **yuan** and includes resources of cement, coal, cotton, grain, etc. In order to feed China's large population, agriculture is the most common occupation. Water buffalo still pull the plows while the women plant the rice seedlings in knee-deep water paddies.

China has been credited with many discoveries and inventions. The silkworm spins cocoons that yield up to 1000 yards of thread on each cocoon. Gunpowder, kites, paper, printing, wok cooking, wheat, the calendar, compass, and The Great Wall have all given China its prestige in the world.

China's history goes back 5000 to 7000 years. Its history is divided into dynasties. Peking, the capital of China, is an interesting city with its elaborate gardens and exclusive emperor's household called the Forbidden City. Ping-Pong has become a major competitive sport. Tai Chi Chaun is an ancient exercise using ballet-type moves. The Chinese New Year, usually celebrated in January, is the most important holiday celebration in the country. Many people travel, work, and live on the water. Large Chinese boats are called junks; small boats, sampans.

Acupuncture and "barefoot" doctors are common forms of medical care. The rest of the world has begun to acknowledge these unconventional forms of care. Family life is also important to the Chinese. The oldest family member does have the most respect. Overpopulation has been a great concern in China. Now the Chinese are encouraged to marry later and plan to have no more than two children. Sons are given more opportunities than daughters, although this is changing in the younger families.

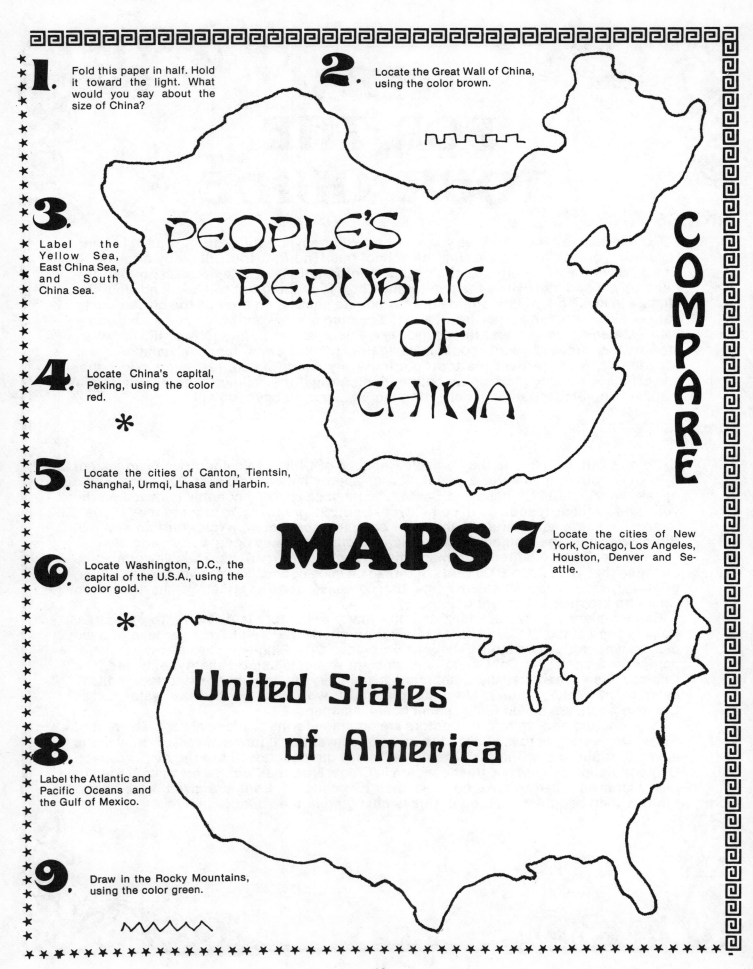

1. Fold this paper in half. Hold it toward the light. What would you say about the size of China?

2. Locate the Great Wall of China, using the color brown.

3. Label the Yellow Sea, East China Sea, and South China Sea.

PEOPLE'S REPUBLIC OF CHINA

COMPARE

4. Locate China's capital, Peking, using the color red.

*

5. Locate the cities of Canton, Tientsin, Shanghai, Urmqi, Lhasa and Harbin.

MAPS

7. Locate the cities of New York, Chicago, Los Angeles, Houston, Denver and Seattle.

6. Locate Washington, D.C., the capital of the U.S.A., using the color gold.

*

United States of America

8. Label the Atlantic and Pacific Oceans and the Gulf of Mexico.

9. Draw in the Rocky Mountains, using the color green.

THE GREAT WALL

The Great Wall of China is the only manmade structure that can be seen by the astronauts on the moon. It was built over 2,000 years ago to protect China from its enemies to the north. Thousands of people worked on the Great Wall making bricks and blocks of granite stone. During its construction many workers died and were buried inside the wall. The wall is over 2,000 miles long. It is wide enough, in some places, for nine soldiers to march side by side. It is also very steep as it winds up the mountains. Towers were built for observation and for shelter. Thousands of people from all over the world go to China to see this great structure that men have made. Many Chinese people also travel great distances to view it.

Below are some questions your tour guide may help you with.

1. How many miles long is the Great Wall?_____
   _____

2. How did the builders cut the rocks to fit together? _____
   _____

3. How did the builders get the rocks up the sides of the mountains? _____
   _____

4. Why do you suppose it can be seen from the moon? _____
   _____

5. What would you like to know about the Great Wall? _____
   _____

Calligraphy

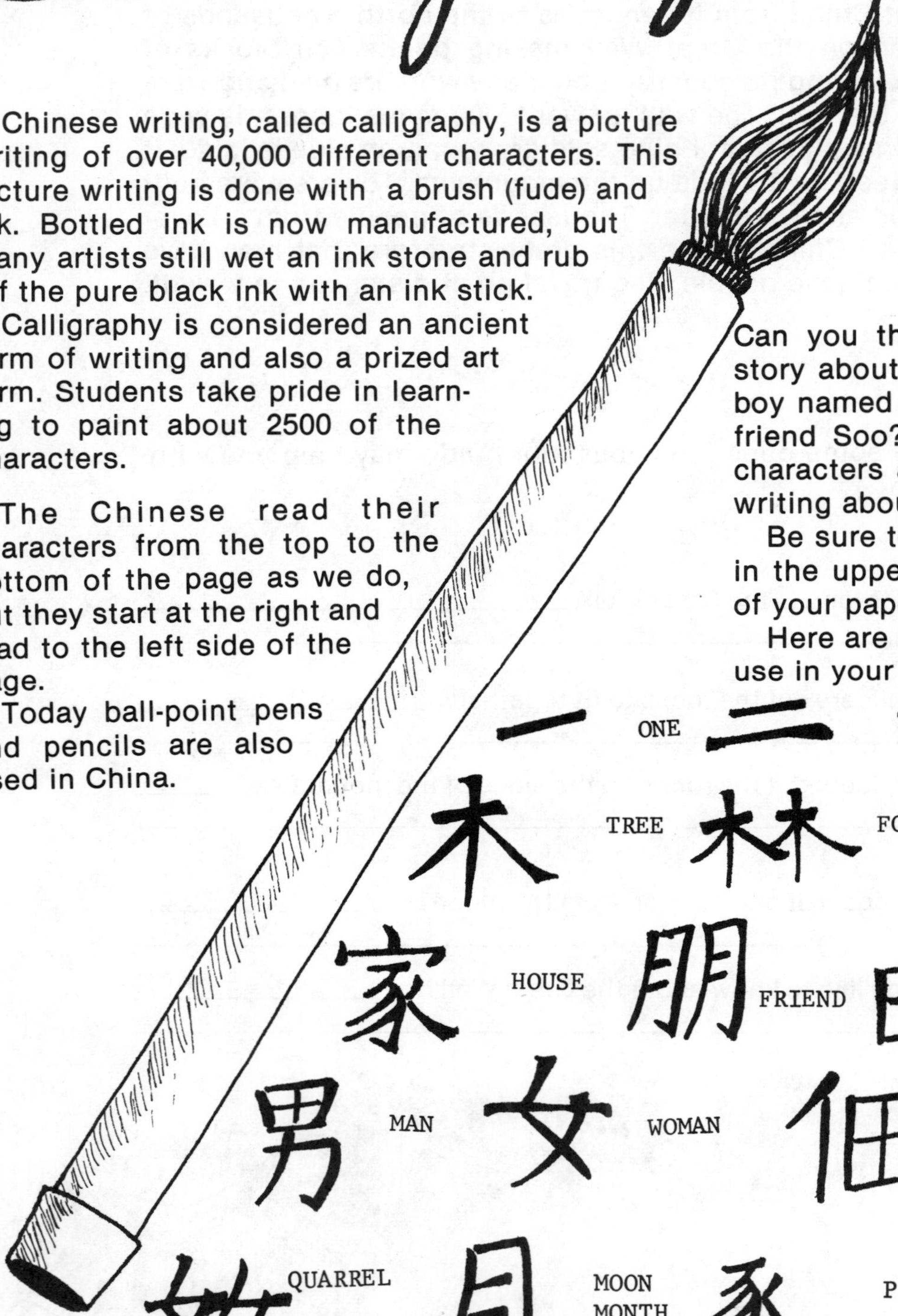

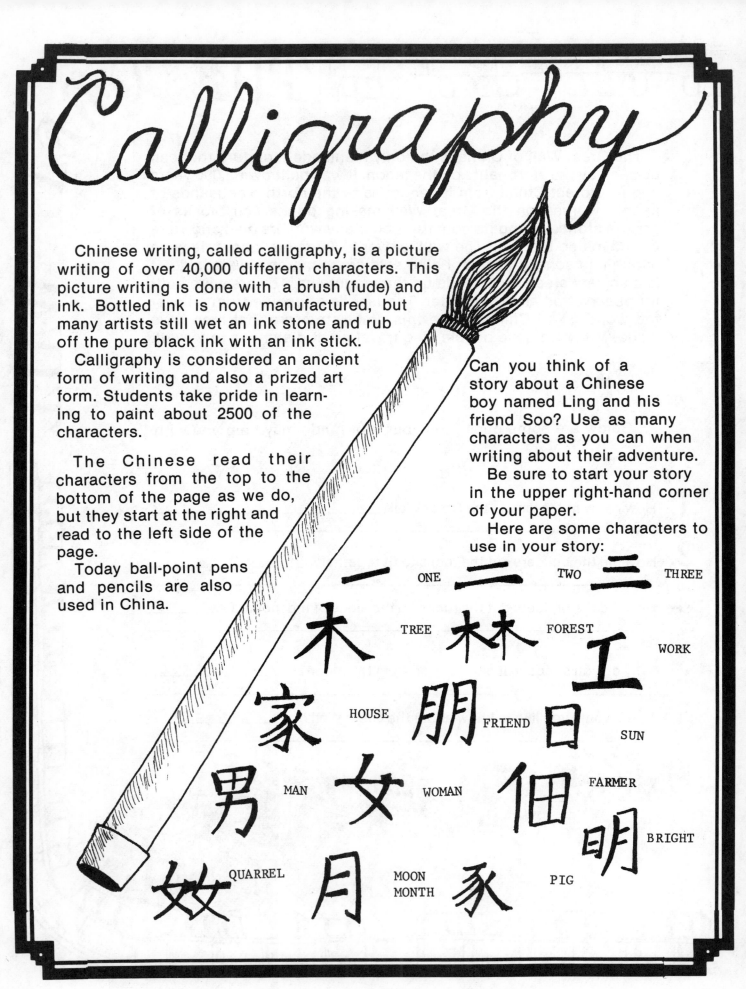

Chinese writing, called calligraphy, is a picture writing of over 40,000 different characters. This picture writing is done with a brush (fude) and ink. Bottled ink is now manufactured, but many artists still wet an ink stone and rub off the pure black ink with an ink stick.

Calligraphy is considered an ancient form of writing and also a prized art form. Students take pride in learning to paint about 2500 of the characters.

The Chinese read their characters from the top to the bottom of the page as we do, but they start at the right and read to the left side of the page.

Today ball-point pens and pencils are also used in China.

Can you think of a story about a Chinese boy named Ling and his friend Soo? Use as many characters as you can when writing about their adventure.

Be sure to start your story in the upper right-hand corner of your paper.

Here are some characters to use in your story:

一 ONE
二 TWO
三 THREE
木 TREE
林 FOREST
工 WORK
家 HOUSE
朋 FRIEND
日 SUN
男 MAN
女 WOMAN
佃 FARMER
明 BRIGHT
奻 QUARREL
月 MOON MONTH
豕 PIG

GIANT PANDAS

The giant panda has a white coat with black arms and legs. Pandas have a black spot around each eye and small black ears. When pandas are fully grown, they may weigh 300 pounds and be six feet tall. They belong to the raccoon family. Can you guess why? _____

_____

Giant pandas live in bamboo forests on the hillsides of Southwest China. Tender bamboo shoots are their favorite food, although they may eat birds and small rodents if bamboo becomes scarce. A thick lining in the pandas' throats protects them from splinters. When they eat, they sit down with their legs stretched out. Sometimes they roll right over on to their backs until their hind legs stick up in the air.

There are only a few more than forty pandas living outside the country of China. These pandas are found in zoos. Can you tell why it is hard to raise pandas in zoos? _____

_____

_____

LIFE ON THE WATER

The Chinese people living near the ocean or large rivers may actually live **on** the water. Houseboats are very common. Many people also travel by water. Junks are large boats. The owners of these large boats often paint eyes on the fronts of them. The owners believe these eyes will help the ship find its way through the water.

Smaller boats are called sampans. These are used for fishing, transportation or housing.

If you were going to live on a boat, what would you want it to look like?

_____

What equipment would you like it to have? _____

_____

Would you want eyes on the front? _____ If so, what would they look like? _____

Draw a picture below of your houseboat. You and your family should be somewhere on your boat.

CHINESE NEW YEAR

The Chinese enjoy celebrating their New Year. It usually occurs late in January or February right after the new moon.

For many days before the celebration the families are preparing food and decorating their homes. During this celebration there is much feasting and visiting of friends and relatives.

A parade of good luck dragons, fireworks, music, and merriment takes place. Dragons are invited into homes and places of business. It is believed they will bring good luck to the people.

Children receive red envelopes with money inside. These envelopes are the children's special gifts.

Everyone has a birthday at this time and becomes one year older—even newborn babies.

How do you celebrate the New Year? _____
_____

Do you have friends or relatives visit? _____

Do you eat special foods? _____

Do you watch a parade? _____

Draw a small picture of what you think the Chinese New Year Parade would look like.

COOKIES

AND COOKING

Chinese cooking is usually done with many of these basic utensils: bamboo steamer, teapot, strainers, wok, and hot pot.

The food is usually cut into bite-sized pieces which can be easily picked up with chopsticks. The food is also cut small so that the cooking time is short and only a little fuel is used.

Everyone enjoys fortune cookies at the end of a tasty Chinese dinner. Make your own fortune cookies with the recipe below. Be sure to write your own fortunes first on strips of paper.

Fortune Cookies

4 egg whites
1 cup sugar
½ cup melted butter
2 tbsp. water

½ cup flour
¼ tsp. salt
½ tsp. vanilla

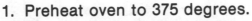

1. Preheat oven to 375 degrees.
2. Blend sugar into egg whites until fluffy.
3. Melt butter and set aside to cool.
4. Add flour, salt, vanilla, water, and butter to the sugar mixture.
5. Beat until smooth.
6. On well-greased cookie sheet, pour batter into 3-inch circles.
7. Bake for 8-10 minutes until golden brown.
8. Lay the fortune paper into the center of cookie. Fold cookie into thirds and bend the center gently.
9. If cookie gets hard to bend, reheat in oven for a minute.
10. Yields 30 fortune cookies.

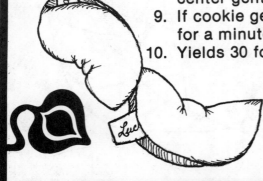

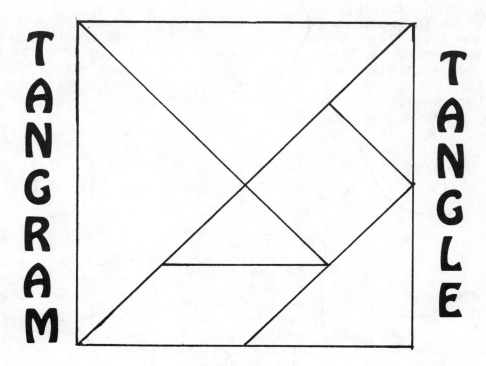

TANGRAM

TANGLE

The tangram puzzle was invented a few thousand years ago by a man named Tan. Trace the puzzle shape onto a square of paper or a square of carpet tile. Carefully cut the pieces apart. Try to form some of the shapes shown below. The real challenge is to get the small pieces back into the tangram's original shape!

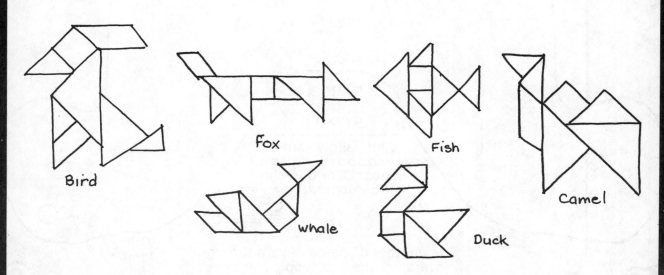

Bird

Fox

Fish

Camel

whale

Duck

Can you form some original shapes to share with your friends?

PAGODA LANTERNS

Many buildings were constructed with graceful roofs of this shape. Much gold was used to decorate these buildings. They were usually shrines honoring someone of importance.

Lanterns, with a candle inside, were originally used for lighting. Now they are used mainly during celebrations.

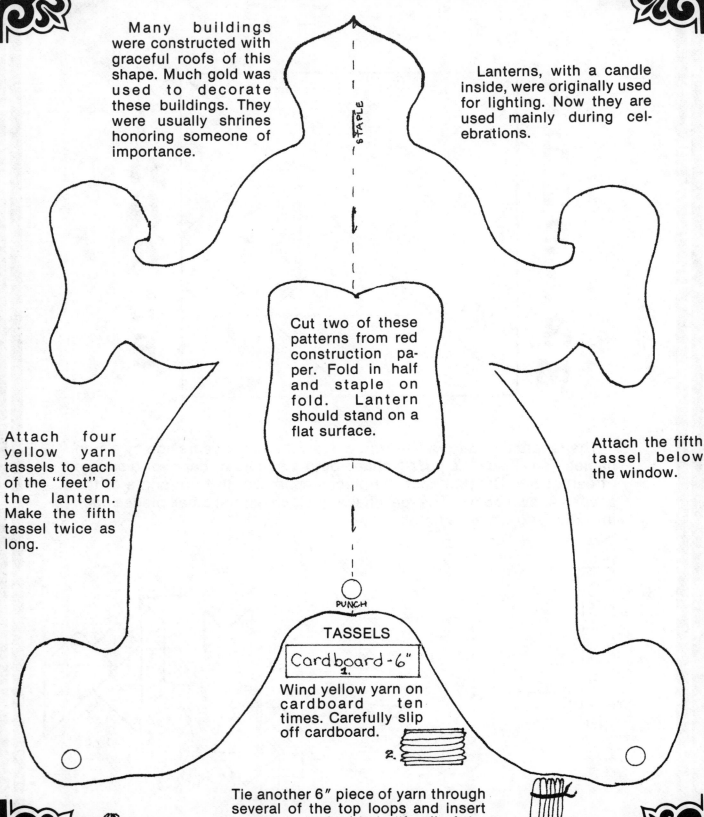

STAPLE

Cut two of these patterns from red construction paper. Fold in half and staple on fold. Lantern should stand on a flat surface.

Attach four yellow yarn tassels to each of the "feet" of the lantern. Make the fifth tassel twice as long.

Attach the fifth tassel below the window.

PUNCH

TASSELS

Cardboard - 6"
1.

Wind yellow yarn on cardboard ten times. Carefully slip off cardboard.

2.

3.

Tie another 6" piece of yarn through several of the top loops and insert into the punched hole "feet" of the pagoda. Repeat for five other tassels. Display by hanging from the ceiling.

4.

THE FOLDING FAN OF FORTUNE

Have a friend choose an animal, and you can tell your friend's fortune!

Animal	Years	Fortune
Ram	1979 1991	Charming
Horse	1978 1990	Popular
Snake	1977 1989	Wise
Dragon	1976 1988	Lucky
Rabbit	1975 1987	Happy
Tiger	1974 1986	Rebellious
Ox	1973 1985	Patient
Rat	1972 1984	Nervous
Pig	1971 1983	Gallant
Dog	1970 1982	Worried
Rooster	1969 1981	Aggressive
Monkey	1968 1980	Mischievous

At the top of each section of this fan, is a list of the twelve animals used in the Chinese calendar. Each animal used has a special meaning. Write a fortune for each of these animals. Use the word in your fortune that describes the animal for that year. Fold the paper into a fan shape and have your friends choose their favorite animals for their fortunes.

CHINA

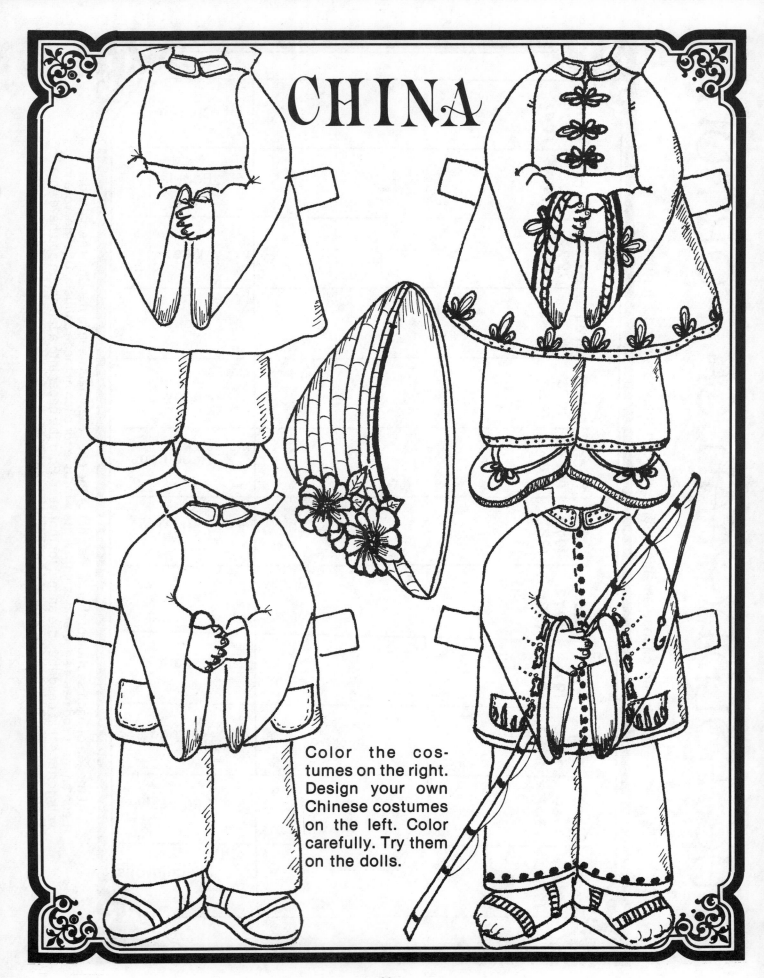

Color the costumes on the right. Design your own Chinese costumes on the left. Color carefully. Try them on the dolls.

TRAVELIN' ON...

The following is a list of additional topics that interested or highly motivated students may want to learn more about:

Inventions: paper, kites, gunpowder, printing, calendar, wheel, compass
The history of the dynasties
Acupuncture points and their effects
The yuan money system
Steps in growing rice
The Forbidden City
Architecture in China
Tai Chi Chaun exercises
Chinese New Year celebrations
Dragons as good luck charms
Chinese opera and ballet
Art forms: ivory carving, cloisonne', cork carving, Oriental rugs, painting, silk, straw pictures, paper cuts
Relationship with Taiwan
Mongals
Tibetans
Musical instruments
Schools
Recipes for Chinese foods

XICO MEXICO MEXICO MEXICO MEXICO
MEXICO
EAT AT CHI-CHI'S
Viva la Mexico
We're Numero Uno!
MEXICO MEXICO MEXICO MEXICO

TIPS FOR THE TOUR GUIDE

Klass Kickoff:

A class activity that will create excitement and make students inquisitive could be an unusual slide show. There are probably many people who have visited Mexico and would have slides you could use. Otherwise, find pictures of Mexico that have no writing on them. Try to include at least one picture of people, food, animals, and houses. This activity will work most successfully if you are the tour guide. Have kids list all the facts they can determine while looking at the slides. An example might be the picture of a person. The clothing would tell the temperature of the climate, the products used to make the clothing, any unusual apparel that's found only in Mexico. From the clothing and temperature you can lead the students to find places on the map that would fall into the temperature range for Mexico. After all slides have been viewed and discussed, lead the students to discover that the country to be studied is Mexico. As mentioned before many people have visited this country, and you may have such a family in your class. Be sure to ask for any information, resources, and classroom presentations that they could help with.

Tips:

There are approximately 65,900,000 people in Mexico. The capital is Mexico City, the largest city in the world. Spanish is the official language. The discovery of oil has boosted the Mexican economy but also caused many economic problems. There is a lack of good farmland and there are too few jobs for unskilled workers. Poverty and riches are evident. There appears to be no middle class. Many families migrate to the United States to find work. This migration has caused many problems in the states of Texas, New Mexico, Arizona, and California. Oil has brought about additional progress in industry and farming. Irrigation is turning deserts into farmland. The Mexican people are fun-loving and put their cares aside when a fiesta is planned.

Mexican markets are a colorful experience with fresh meats, fresh fruits, vegetables, household items, clothing, jewelry, etc., on display. Bartering is expected and enjoyed by the merchants! The currency standard is the **peso**.

Because of the long ranges of mountains in both the east and west, the climates range from tropical on the coastlines to cactus deserts in the central highlands.

The history of Mexico is colorful, interesting and often inspiring.

NORTH AMERICA

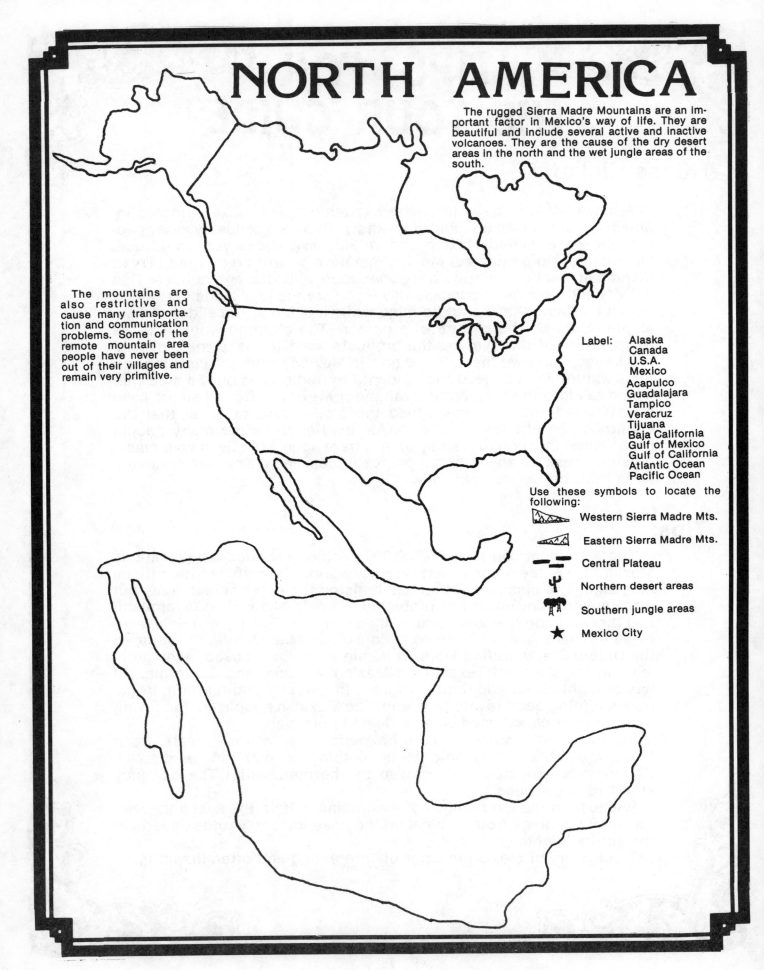

The rugged Sierra Madre Mountains are an important factor in Mexico's way of life. They are beautiful and include several active and inactive volcanoes. They are the cause of the dry desert areas in the north and the wet jungle areas of the south.

The mountains are also restrictive and cause many transportation and communication problems. Some of the remote mountain area people have never been out of their villages and remain very primitive.

Label: Alaska
          Canada
          U.S.A.
          Mexico
          Acapulco
          Guadalajara
          Tampico
          Veracruz
          Tijuana
          Baja California
          Gulf of Mexico
          Gulf of California
          Atlantic Ocean
          Pacific Ocean

Use these symbols to locate the following:

Western Sierra Madre Mts.

Eastern Sierra Madre Mts.

Central Plateau

Northern desert areas

Southern jungle areas

★   Mexico City

26

FIESTAS

There are two kinds of festivals in Mexico: political and religious. Mexicans celebrate many of their holidays with colorful fiestas.

One of the biggest celebrations occurs on September 15 and 16. This is for Independence Day! Find out more about the reason for this celebration. Is it similar to our Fourth of July festivities? Write your findings on the back of this paper.

Most of the Mexican people are Roman Catholics and celebrate the birth of Christ from December 12th through January 6th. The **posada**, or reenactment of Mary and Joseph looking for a place to stay, is portrayed for several nights before Christmas. Then on Christmas Eve, the actors are invited into the homes of friends where a party takes place. The pinata is an important part of the celebration for children. Festivities and parties continue until January 6th when gifts are exchanged. This is Epiphany when the wise men arrived in Bethlehem.

Read more about the posada and try to act out this little play. Perhaps you could end with a pinata party.

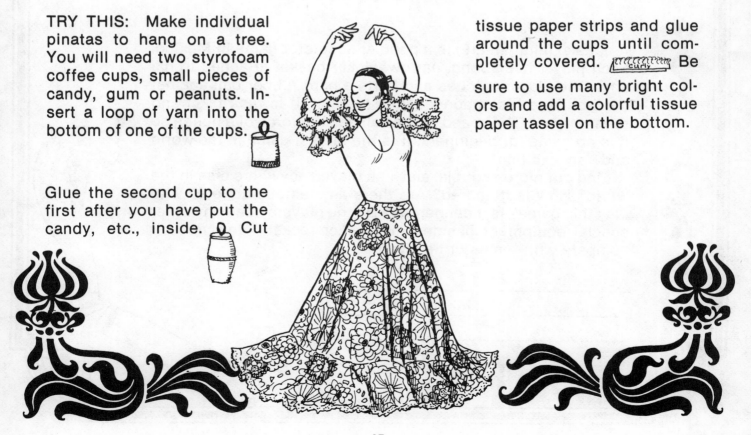

TRY THIS: Make individual pinatas to hang on a tree. You will need two styrofoam coffee cups, small pieces of candy, gum or peanuts. Insert a loop of yarn into the bottom of one of the cups.

Glue the second cup to the first after you have put the candy, etc., inside. Cut tissue paper strips and glue around the cups until completely covered. Be sure to use many bright colors and add a colorful tissue paper tassel on the bottom.

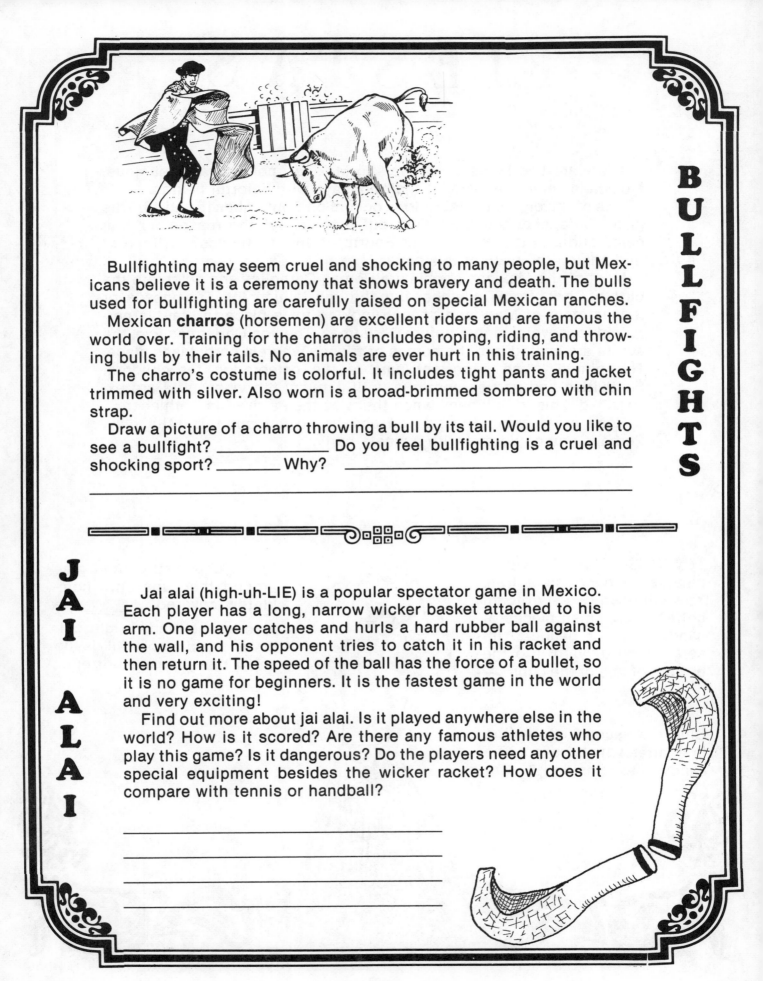

Bullfighting may seem cruel and shocking to many people, but Mexicans believe it is a ceremony that shows bravery and death. The bulls used for bullfighting are carefully raised on special Mexican ranches.

Mexican **charros** (horsemen) are excellent riders and are famous the world over. Training for the charros includes roping, riding, and throwing bulls by their tails. No animals are ever hurt in this training.

The charro's costume is colorful. It includes tight pants and jacket trimmed with silver. Also worn is a broad-brimmed sombrero with chin strap.

Draw a picture of a charro throwing a bull by its tail. Would you like to see a bullfight? _____ Do you feel bullfighting is a cruel and shocking sport? _____ Why? _____

_____

Jai alai (high-uh-LIE) is a popular spectator game in Mexico. Each player has a long, narrow wicker basket attached to his arm. One player catches and hurls a hard rubber ball against the wall, and his opponent tries to catch it in his racket and then return it. The speed of the ball has the force of a bullet, so it is no game for beginners. It is the fastest game in the world and very exciting!

Find out more about jai alai. Is it played anywhere else in the world? How is it scored? Are there any famous athletes who play this game? Is it dangerous? Do the players need any other special equipment besides the wicker racket? How does it compare with tennis or handball?

_____

_____

_____

MEXICO'S

1.

25 centavos

You buy 2 pineapples for a party. How much do you spend? _____

2.

50 centavos

You give the shopkeeper 1 peso. How much change do you get back? _____

3.
84 centavos

Pinatas are so much fun! The children will need 3 of them—all different shapes. How much do you spend? _____pesos _____centavos

6.

3 centavos

Tomatoes for your tacos. They are so red! You need 4 for your recipe. How much did you spend? _____

4.

2 pesos

Aah! a beautiful serape for you. Will you have enough money? You have 1 peso and 100 centavos. _____

5.

16 centavos

A new clay cooking pot! What a pretty design. You will buy 2. How much did you spend? _____

7.

4 pesos

This piglet will grow to be worth 5 times as much as now. How much can you sell it for later? _____

8.
63 centavos

A whole basket of apples for only 63 centavos! What a buy! You take 2 baskets for your 8 children. How much did you spend? _____pesos _____centavos

9.

17 centavos

Wouldn't these crepe paper flowers add color to your adobe home? Buy 5 brightly colored ones for _____.

The currency used in Mexico is the **peso**. It takes 100 **centavos** to make 1 **peso**.

MAGICAL MARKET

tissue paper flowers

Cut four petal pieces from tissue paper. Place one on a flat surface. Cut the stamen, etc., from yellow or black tissue paper. Roll into a bundle. Place in the middle of the first petal. Lay the second petal on top. Place the third and fourth petals on each side. Pinch the center of all four flower petals. Twist the center *tightly* with fine wire or a pipe cleaner. Then fold bottom petals carefully to top.

Add green leaves if you wish.

These flowers can be used for decorations on bulletin boards. Attach a long green stem and place in vases. They are perfect to decorate a stage for a program. Girls could wear them in their hair.

Using 8″ x 11″ flat slabs of terra cotta clay, make an outdoor market scene. Design the market buildings first, adding much detailing with the rough wooden planks and the windows of the buildings. Cut the clay to fit the outline of the marketplace. Then make small pieces of fruit for the fruit stand; clay jars and pots for the pottery; clay baskets for the straw area; hanging pinatas, serapes, etc. Wet the clay and press the small pieces in place. Check with your art teacher about firing the flat pieces. Don't forget a hole at the top if you plan to hang them!

After firing, paint with acrylic paints. Accentuate the details.

OR: Use regular modeling clay, small sticks and a cake or jelly roll pan of sand. Construct the market buildings with small sticks. Make small items for sale out of the modeling clay. Don't forget some people, too! Place these clay items on the pan with sand. You could add cacti and any other items for interest.

CLAY MARKETPLACES

SAUCEY SOMBRERO AND SNAPPY GUITAR

A Mexican fiesta is always fun to have. You may consider inviting other classes and/or parents to your fiesta.

Some suggested activities might include:

"Mexican Hat Dance"

A short play about the Christmas **posada**

A bullfight skit

Breaking a pinata

A tasting party

A display of the map work and any other research

Costumes are always fun to make. A pattern for Mexican sombreros follows. You may want to wear a white blouse or shirt and then tie long crepe paper sashes around your waist. (Be sure to use brightly colored crepe paper!)

Old pillowcases with holes cut for heads and arms make great serapes. Decorate with fabric crayons or magic markers. (Be sure to put folded newspaper inside so the markers do not bleed through.)

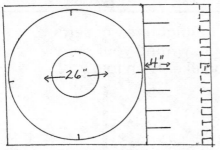
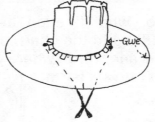

Decorate the brim. Attach a cord as a hat band. The hat does not need to be worn on the head, but can be dropped over the shoulders and held in place with the cord.

Guitars can be cut out of corregated cardboard using yarn or string for the strings. Decorate and add a strap so the guitar can be worn, also.

Try writing a travel brochure to inform tourists as to what they will see when traveling to Mexico. Use the plan below to help you write.

3 What kind of_? Which_?	1 Who? What?	2 What happened? What do they do?	4 What, where, how, why?
Example: many, some all, few	Mexicans	are rich like to sing like bullfights	the wages are low with guitars for the excitement
	Fiestas		
	Homes		
	Christmas		
	Tacos		

Add more "Who, what?" such as bullfighters, deserts, jungles, etc. After filling in the chart, begin writing your story. Add more sentences and words where you like. Then fold a 12″ x 18″ piece of paper into six parts.

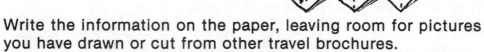

Write the information on the paper, leaving room for pictures you have drawn or cut from other travel brochures.

You could also use as a cover to an invitation to your Mexican fiesta program.

TRAVELIN' TOURIST GUIDE

Here Comes the Sun

The homes of most of the peasant people are often made from tree branches that are held together with mud chinking. The mud is pressed into the holes between the branches.

The roofs are usually made of grass or palm leaves. They are called **thatched** roofs.

The sun's rays can be very hot in this land so near the equator. To show the amount of heat given off by the sun's rays, try these experiments on a sunny day.

1. Place a thermometer in a sunny place and cover and record the temperature on the sun's face (above). Do the same experiment, only this time place the thermometer in the shade. Record on the cloud shape.
2. Do the above experiment again replacing the grass with white paper. Read and record the temperatures.
3. Use black paper this time. Read and record the temperatures.

Then discuss the importance of the color of the house roofs in your city. You may need to take a short walk in the neighborhood and record the color of the roofs.

SPANISH SPEAKO

To the teacher: Make "meaning" cards for each of the words listed above. These are the cards that are drawn for the game.

Use the Spanish words below for your Speako game. Copy one word in each box above. Choose words at random; do not copy them in order. There will be words left over.

Cut out your Speako card. Then play with a group of your friends. Use the rules that apply to Bingo.

libro-book	manana-tomorrow	mesa-table	muchacho-boy
musica-music	noche-night	papel-paper	paz-peace
puerta-door	rio-river	rosa-rose	senorita-lady
aeroplano-airplane	amigo-friend	bandera-flag	bella-beautiful
burro-donkey	cafe-coffee	casa-house	comprende-understand
sombrero-hat	dulce-candy	ensalada-salad	escuela-school
flor-flower	gato-cat	gracias-thanks	hacienda-farm

Have a tasting party!

Cook in class, ask parents to cook or contribute ingredients, or enlist a nearby Mexican restaurant to help!

It is not necessary for each child to have a large portion of any one food. Small, tasting pieces are sufficient!

Be sure to get recipes from parents or restaurants. Make a Mexican cookbook. Children could design a large taco (with all of the goodies inside) for the cover.

You might also want to include a few positive candid remarks by the children about each dish.

A recipe for tacos follows. This would be for a class of twenty-five students.

Tacos

Make tortillas by rolling refrigerator biscuits (in the tube can) with a rolling pin or the side of a tin can. Be sure to have flour or cornmeal handy in case the biscuits stick.

Deep-fat fry these flat "tortillas." This does not take long. Drain on paper toweling and bend in half while still warm.

The filling is:

2 lbs. ground beef (browned and drained)
2 envelopes of taco filling mix
(follow the instructions on the package)
1 lb. cheddar cheese (shredded)
3 large tomatoes (chopped)
1 large head lettuce (chopped)

Additional ingredients might include Tabasco sauce, sour cream, onions, green peppers, olives, Parmesan cheese, or any other favorites.

Enjoy!

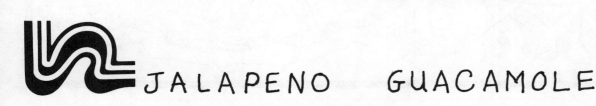

TORTILLAS NACHOS CHIMICHANGAS

TOSTADA REFRIED BEANS TOSTADA

JALAPENO GUACAMOLE

Color the costumes on the right.

Design your own Mexican costumes on the left. Color carefully.

TRAVELIN' ON...

The following is a list of additional topics that interested or highly motivated students may want to learn more about:

Volcanoes
Oil production
Meaning of the flag emblem
Wetbacks
1910 Peasant's Revolt
Music of Mexico
1810 Rebellion
Pyramids at Teotihuacan
Aztecs
Influence of gold
Hernando Cortes
Learn more Spanish
Major products
Temperature graph for five Mexican cities for all twelve months
Mexican bark painting
Tree of life
Tin Christmas ornaments
God's Eyes (Ojos de Dios)
Molas
Silver (Taxco)
Acapulco
Mayan Indians (Yucatan)
Hieroglyphics

AUSTRALIA

Australia Jumps

DON'T
FEED
THE
POUCHES

TIPS FOR THE TOUR GUIDE

Klass Kickoff:

Set up a *huge* gameboard on the floor of your classroom. Use 12″ x 18″ pieces of construction paper as the steps or the track to follow. On several pieces write such things as: kangaroo leap, go ahead 3 spaces; jeep breaks down, go back 2 spaces; strong desert winds, go back 3 spaces, etc. Cover a large, square corrugated cardboard box with plain paper. This will be the die. Write the number or put the appropriate number of dots on each side. Have students roll the *huge* die. Then move the designated number of spaces. Whoever reaches *home* first is the winner. If you use this game again at the end of the study, have the students answer a question pertaining to Australia in order to move the number on the die. The important thing is to have fun!

Tips:

It is believed by some that Australia broke off from a larger land mass millions of years ago. After this happened, the animals on this island evolved in unusual ways. This could be the reason there are so many unusual animals in Australia.

Australia was settled by Britain about 200 years ago. They used the island as a place to send prisoners. Now Australia is much like all countries with large, modern cities, suburbs and countrysides. Most of the population lives on the eastern shore of Australia. West of the Great Dividing Range it is dry, dusty and a desert. This area is called the *outback.* Sheep ranching is a major industry, and Australia is the world's leading producer of wool. Dingos, or wild dogs, are a menace to the ranches. The world's longest fence (6000 miles) was built to protect the sheep.

The population of the Commonwealth of Australia is approximately 14,227,900 dispersed on 2,970,000 square miles of land. The capital of the five "states" is Canberra. The official language is English, and the unit of currency is the dollar. Some of the important industries are coal, dairy products, fruit, grain, iron, machinery, meat, natural gas, oil, silver and sugar.

An easy-to-read reference book for children is **Take a Trip to Australia,** by David Truby. The publisher is Franklin Watts Limited in London.

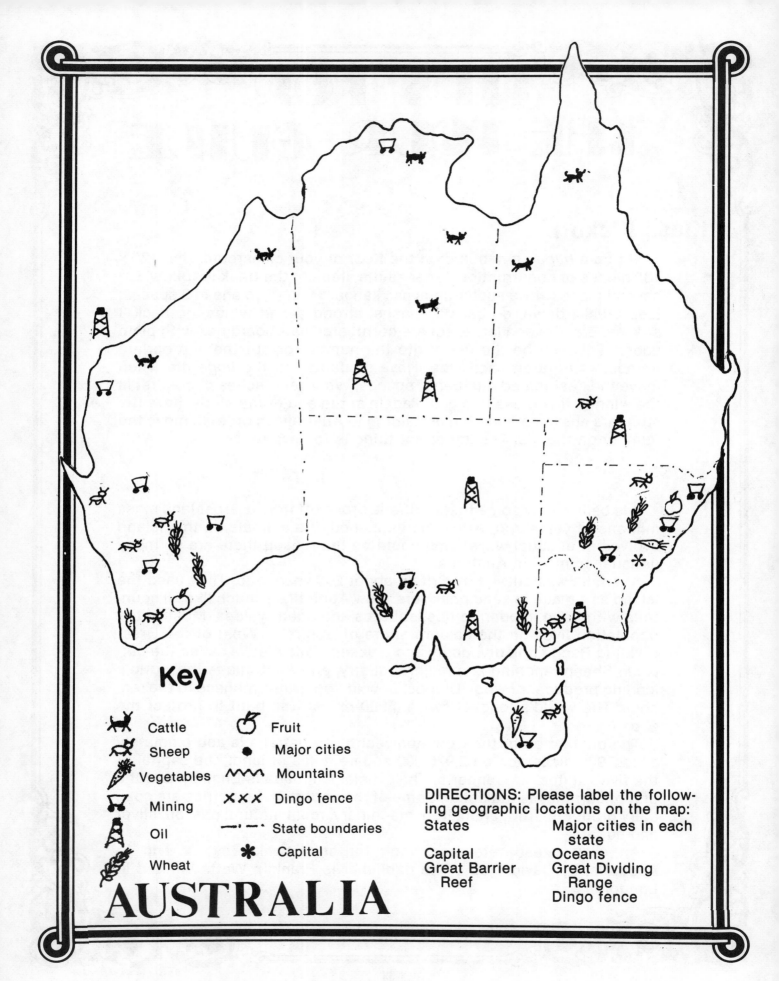

Key

🐕 Cattle

🐑 Sheep

🥕 Vegetables

🛒 Mining

⛽ Oil

🌾 Wheat

🍎 Fruit

● Major cities

∿∿∿ Mountains

✕✕✕ Dingo fence

—··— State boundaries

✳ Capital

DIRECTIONS: Please label the following geographic locations on the map:

States   Major cities in each state

Capital   Oceans

Great Barrier Reef   Great Dividing Range

Dingo fence

AUSTRALIA

COMMONWEALTH OF AUSTRALIA

The British flag appears in the upper left-hand corner of the Australian flag. Under the miniature British flag is the seven pointed white Commonwealth Star. The five white stars on the blue field represent the constellation of the Southern Cross. Can you answer these questions?

1. Why would Australia have a small British flag on its country's flag? _____
_____

2. Why would the constellation of the Southern Cross be important to the Australians? _____
_____

Australia is located in the Southern Hemisphere; therefore the seasons are reversed from our own. Try to list the months when these seasons occur in Australia.

SUMMER _____ TO_____
FALL _____ TO_____
WINTER _____ TO_____
SPRING_____ TO_____

Koala bears spend most of their lives in trees. They nibble eucalyptus leaves which are just about the only food they eat. They do eat about two and a half pounds of leaves per day but drink very little water. The moisture in these leaves supplies their water.

Koalas are marsupials and carry their young in pouches. The newborn koala is no larger than a nickel and must blindly find its way into its mother's pouch. There it lives for several months. When it outgrows the pouch, it rides on its mother's back.

The paw of the koala bear is unusual. Can you guess why it is shaped like this?

_____
_____
_____
_____.

KOALAS

KOALAS

KOALAS

KOALAS

THE INCREDIBLE "shrinking" KOALA

MATERIALS NEEDED:

styrofoam meat tray
    OR
styrofoam picnic plate
permanent magic markers
paper punch
scissors

brightly colored cord or yarn
Teflon cookie sheet
    OR
aluminum foil-covered cookie
   sheet

DIRECTIONS: Preheat oven to 300 degrees. Cut a styrofoam meat tray or a styrofoam picnic plate into a circle. Draw your own koala bear in the center of the circle. Color the koala with permanent magic markers. Outline the koala with a black line. Add the words: I ♡ Koalas! Punch a hole at the top of the circle if you plan to wear it, or use this for a key chain or a zipper pull.

Place the styrofoam circle on a cookie sheet. (See materials list above.) *Watch it carefully as it heats. It will twist and turn and shrink!*

Shrink !

When it is smaller and relatively flat, take it out of the oven. Let it cool.

Measure the length you need to wear around your neck. Add six inches for the knot or bow. Use string, brightly colored cord or fishline to string the panda for a pendant, a key chain, or a zipper pull for your jacket.

One of the greatest tourist attractions of Australia is the Great Barrier Reef. The colorful coral reef stretches 1250 miles along the northeast coast. Many species of fish live in this area where food and shelter are plentiful. The temperature of the water remains a constant 65 degrees in order for the **polyps** (tiny marine animals) to grow and produce coral.

Many types of unusual fish can be found in this area. The queen angelfish is a favorite of tropical fish fanciers and the trumpet fish stalks its prey by patiently standing on its head.

There are many types of coral found in the weed beds. The parrot fish chews the sharp coral to get the tiny polyps out of it for food. Then the parrot fish spits out the coral sand pebbles. This fish is responsible for producing many miles of coral sand reefs.

See if you can find pictures of these corals. Draw a picture of each one. Can you tell how each got its name?

TYPES OF CORAL

Tubeworm   **Brain**

Sponge   FAN

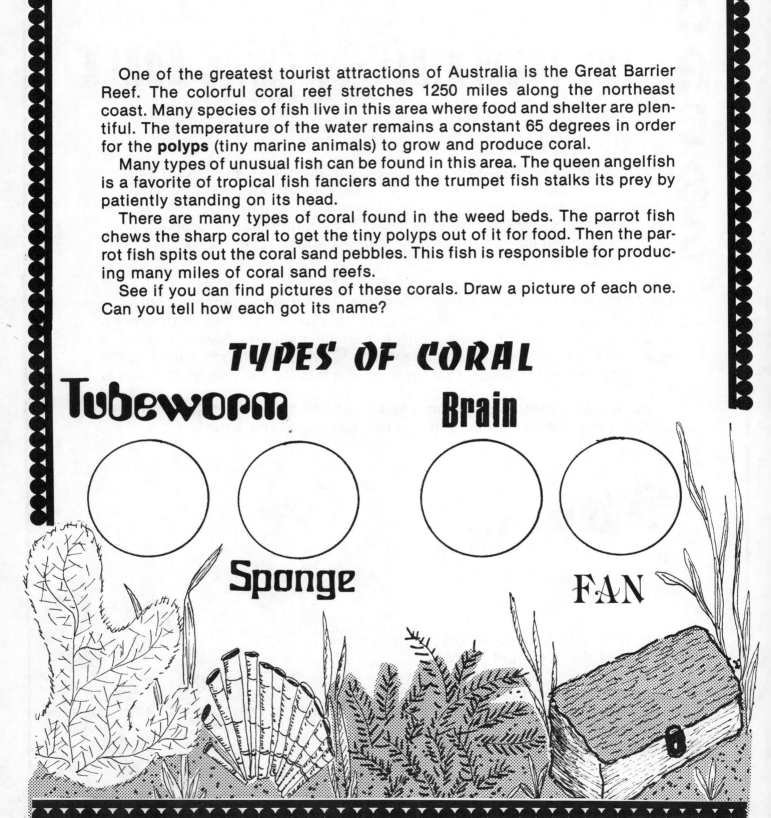

Barrier Reef

Polyps are the tiny, often microscopic marine life that builds the complex coral for its home.

Label the four main parts of this polyp. Use these words: mouth, limestone cup, stomach, tentacles.

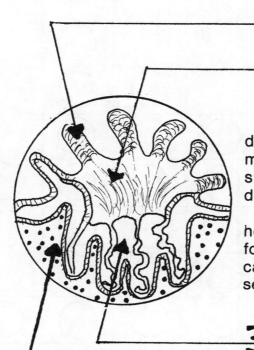

The food or plankton is found in the water with the use of tentacles. When the polyps are frightened or resting, they pull the tentacles inside of the crusty limestone shell.

The food is then deposited into the mouth and moved to the stomach area to be digested.

The stony cup is the home of the coral. It is formed from the calcium carbonate found in the sea water.

It is amazing to think that these microscopic marine animals are capable of building huge reefs and even islands composed mainly of coral.

Kangaroo
Kilometer Contest

km	Karen	Keith	Kevin	Kerrie	Karol	Kay	Kathy	Kelly
40								
35					○			
30				○	○			
25			○	○	○			○
20	○	○	○	○	○	○		○
15		○	○	○	○	○	○	○
10	○	○	○	○	○	○	○	○
5	○	○	○	○	○	○	○	○

km

1. Which kangaroo won the kilometer contest? _____
2. Which kangaroo came in last? _____
3. Which two kangaroos tied? _____
4. How much farther did Kerrie go than Karen? _____
5. How far did Kathy and Keith jump together? _____
6. If Kay was allowed to double her distance, how far would she have jumped? __
7. If Kerrie had stopped halfway, how far would she have jumped? _____
8. How much farther did Karol go than Kathy? _____
9. How far did Karol and Kathy jump together? _____

1. Karol 2. Karen 3. Kevin, Kelly; Keith, Kay 4. 20 km 5. 35 km 6. 40 km 7. 15 km 8. 20 km 9. 50 km

46

The Outback

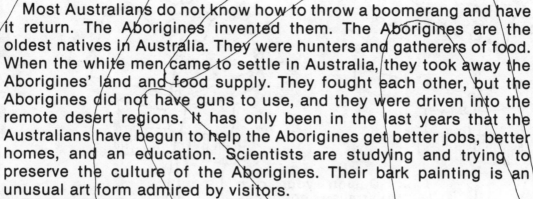

Most Australians do not know how to throw a boomerang and have it return. The Aborigines invented them. The Aborigines are the oldest natives in Australia. They were hunters and gatherers of food. When the white men came to settle in Australia, they took away the Aborigines' land and food supply. They fought each other, but the Aborigines did not have guns to use, and they were driven into the remote desert regions. It has only been in the last years that the Australians have begun to help the Aborigines get better jobs, better homes, and an education. Scientists are studying and trying to preserve the culture of the Aborigines. Their bark painting is an unusual art form admired by visitors.

What do you think?

1. How do you think boomerangs were invented?_____
    _____

2. Does this story about the Aborigines' history remind you of a group of American people? Who? What happened? _____
    _____

3. Why is it important for the Aborigines to get an education?_____
    _____

Try This:

Make cardboard copies of these boomerang shapes. Practice flying them. Which shape seems to fly the farthest? _____
Does any one of them turn around or curve to come back to you?_____ Have a contest! Determine the distance your boomerangs flew. Whose flew the farthest? Graph the results.

The Aborigines

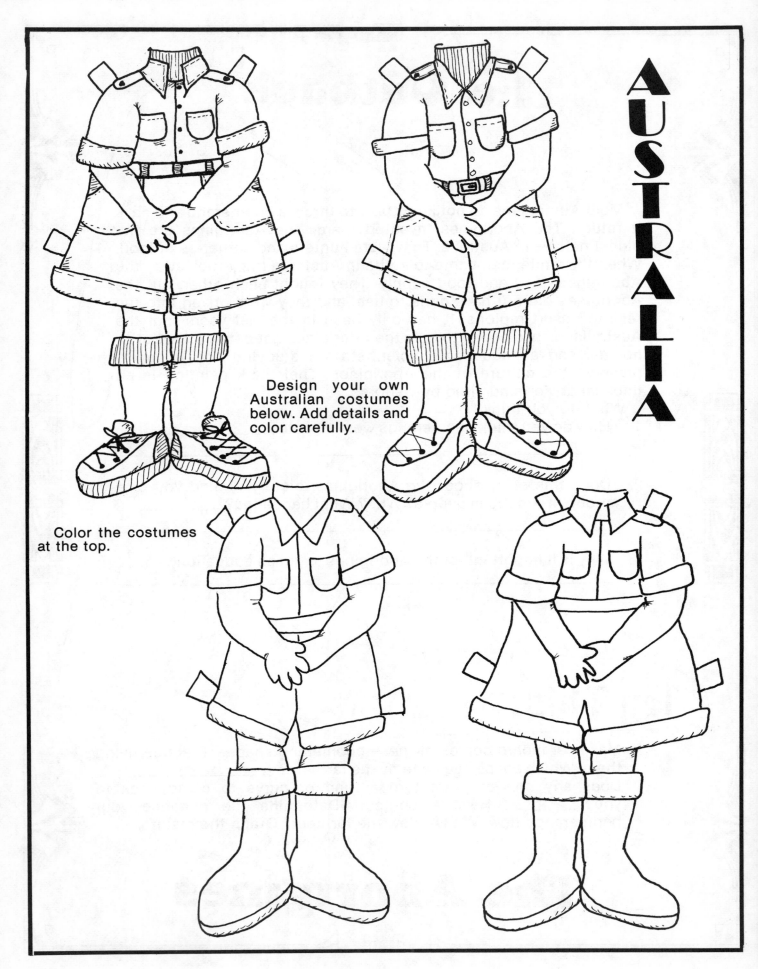

Design your own Australian costumes below. Add details and color carefully.

Color the costumes at the top.

AUSTRALIA

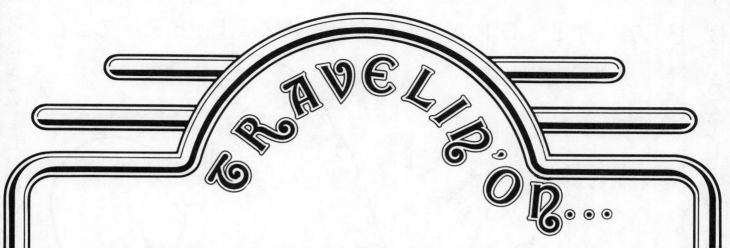

Travelin' On...

The following is a list of additional topics that interested or highly motivated students may want to learn more about:

Kookaburra
Wombat
Duck-billed platypus
Dingo
Spiny anteaters—echidna
Emu
Lyrebird
Tasmanian devil
National crest and the meaning of its symbols
Eucalyptus tree
The wattle flower
Captain James Cook
Sheep ranching
Careers
Mining
1850's gold discovery
Famous singers or sports figures who came from Australia
Tasmanian history
Population
Importance of the Snowy River Scheme
Game of Cricket
Australian rules for football vs. NFL rules for football
Corroboree celebrations

BRAZIL

PIRAN
MAKE GOOD
PETS!

I
LOVE
CHOCOLATE

TIPS FOR
THE TOUR GUIDE

Klass Kickoff:

To draw attention to the country of Brazil, bring a large suitcase into the classroom. Inside, have samples of products from Brazil, such as the following: a chocolate candy bar, cocoa, Brazil nuts, rubber ball, stuffed parrot, butterfly, empty aquarium (piranha), swimmingsuit, plant, section of a cane fishing pole (blowgun), boat, box of tapioca, artifical flower, mahogany, bow and arrow, stuffed monkey, snake, coffee, etc. Unpack the suitcase slowly, showing each object. You can discuss each item's use and origin. Hopefully, the students will determine the climatic area where these objects could be found. Then locate those hot climates on a world map and have the students then vote to see which country of the world you will study next. With your expert guidance, they should be led to the country of Brazil.

Tips:

The Amazon River runs through the center of the Amazon jungle. This river is one of the largest rivers in the world.

There are no seasons, such as fall, winter, spring, etc., but rather the wet season and the dry season. Heavy rains fall each day during the wet season. The dry season has rain only every few days. It does rain year round. Because of its location near the equator, the temperatures are constantly high.

The jungle floor is covered with green plants and vines that twine their way up the tall trees toward the light. Growth is thick and makes travel through the jungle area very difficult.

There are many fruits, berries, nuts and roots to use for food in the jungle. There are also many pigs called tapirs, three-toed sloths, jaguars, armadillos and even anteaters. Birds are plentiful and colorful. There are many alligators and poisonous snakes, such as the anaconda, that live along the riverbanks.

Insects are very numerous and thrive in this warm, damp climate. There are many flying insects and also on the ground can be found spiders, ants, scorpions and bugs.

Fish are plentiful in the ocean and also the Amazon River. These are an important source of food.

Most people live in the cities along the eastern shores of Brazil. The cities are modern and have all the usual conveniences. The Indians who live in the jungle areas live many days journey from any city. Most will never leave their own villages during their entire lives.

Brazil is an interesting country. It has large amounts of land, large populations but so much unpopulated land. It will be interesting to see if the government is able to draw the people to the interior of the country.

BRAZIL

LOCATE
Brazil
Amazon River
Brasilia
Atlantic Ocean
Pacific Ocean
São Paulo
Rio de Janeiro
Rio Branco
Equator

key

Brasilia was built in the state of Goiás. It represents the goal of the country to develop the land away from the Atlantic Ocean. Most of the big cities are seaports located on the coast.

The city is very modern and has unique architectural designs. The government building is located in the shape of a triangle called "Plaza of the Three Powers." The triangle is to symbolize equilibrium maintained by the three branches of government. They are the Congressional Palace, the Palace, the Palace of Justice, and the Executive Palace.

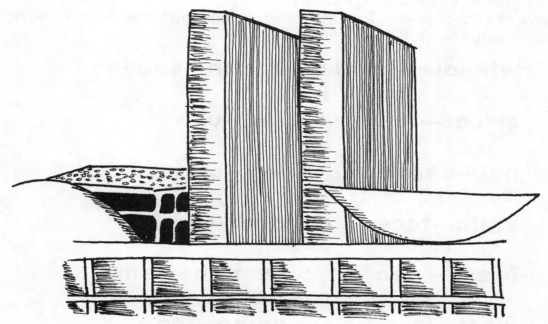

Can you design a modern school for the school of Brasilia?

BRASILIA

PORTUGUESE
The Native Language

Brazil is the largest Portuguese-speaking country in the world. Its common language unites this big country.

Below is a list of several Portuguese words and their meanings. Try to fit the words into the story.

window—janela   health—saúde

street—rua   the—o, a

hat—chapéu   leg—perna

knife—faca   door—porta

bread—pao   men—homens

wool—la   game—jogo

are—sao   women—senhoras

sky—céu   church—igreja

One early morning when the _____ was clear and bright, a boy named Pedro walked down the _____ of the little town. He knew all of the _____ and _____ would be in _____ on this Sunday. He peered into the _____ of his small house. He hurried through the _____ of the house, took the _____ from the small table and cut a warm slice of bread. Oh, how good it tasted!

Where had Pedro been all night? Would his mother care that he had taken the **pao** or was she saving it for a special treat? Can you finish this story?

_____

_____

_____

To the teacher: Hand out a chocolate kiss or small square of chocolate to each child. Then have the children do these activities. (Cover these instructions before you duplicate.)

Smell the chocolate. Think of five words that describe how it smells.

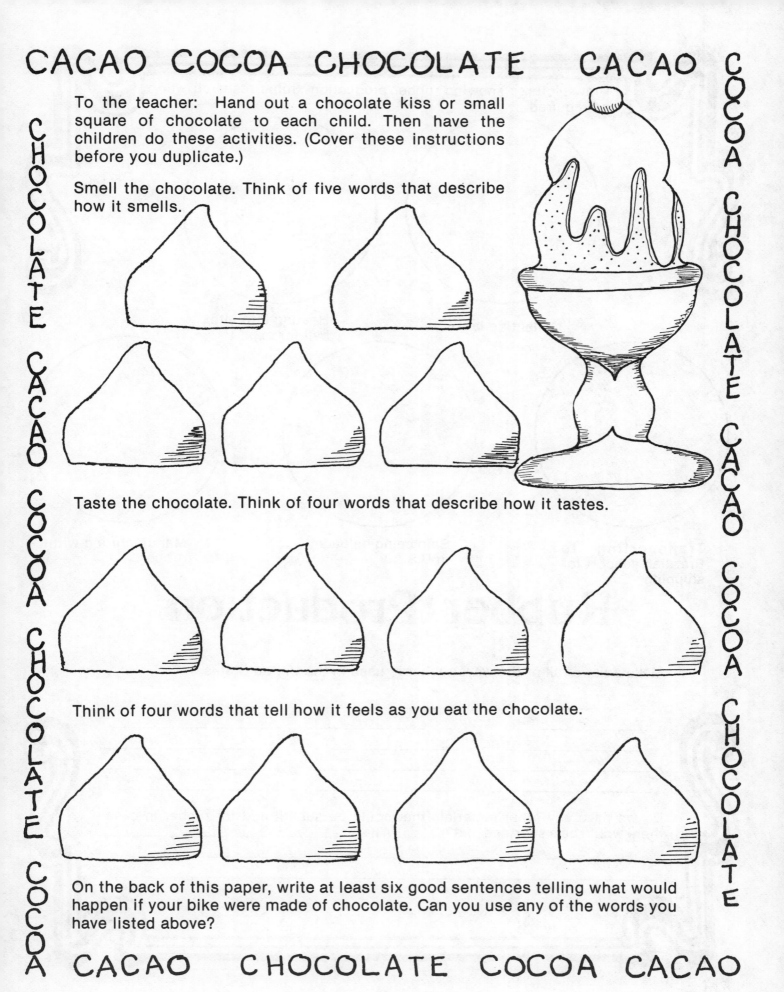

Taste the chocolate. Think of four words that describe how it tastes.

Think of four words that tell how it feels as you eat the chocolate.

On the back of this paper, write at least six good sentences telling what would happen if your bike were made of chocolate. Can you use any of the words you have listed above?

Draw pictures showing rubber production. Some research will be required.

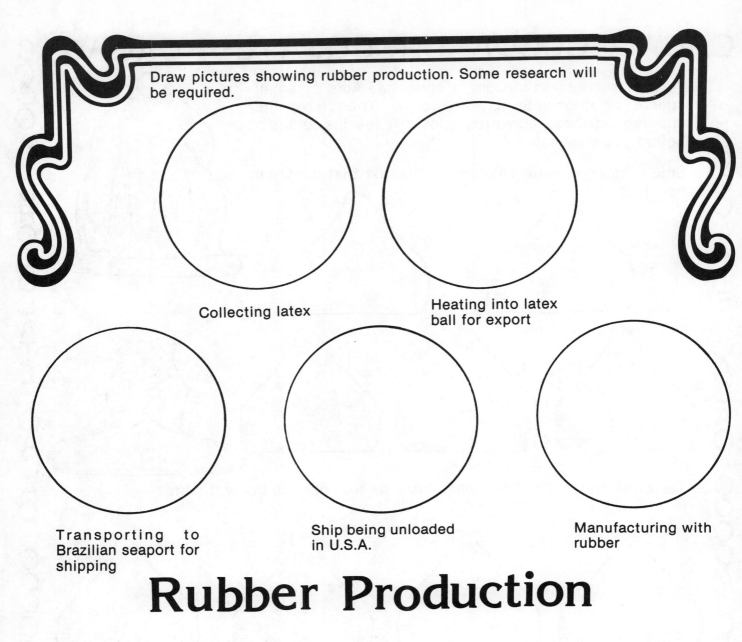

Collecting latex

Heating into latex ball for export

Transporting to Brazilian seaport for shipping

Ship being unloaded in U.S.A.

Manufacturing with rubber

Rubber Production

Make a list of all of the ways rubber is used in the United States.

_____   _____   _____
_____   _____   _____
_____   _____   _____
_____   _____   _____

Do we have any other materials that could be substituted for rubber in case there would be a shortage of it? _____ What? _____
_____
_____

JUNGLE ANIMALS

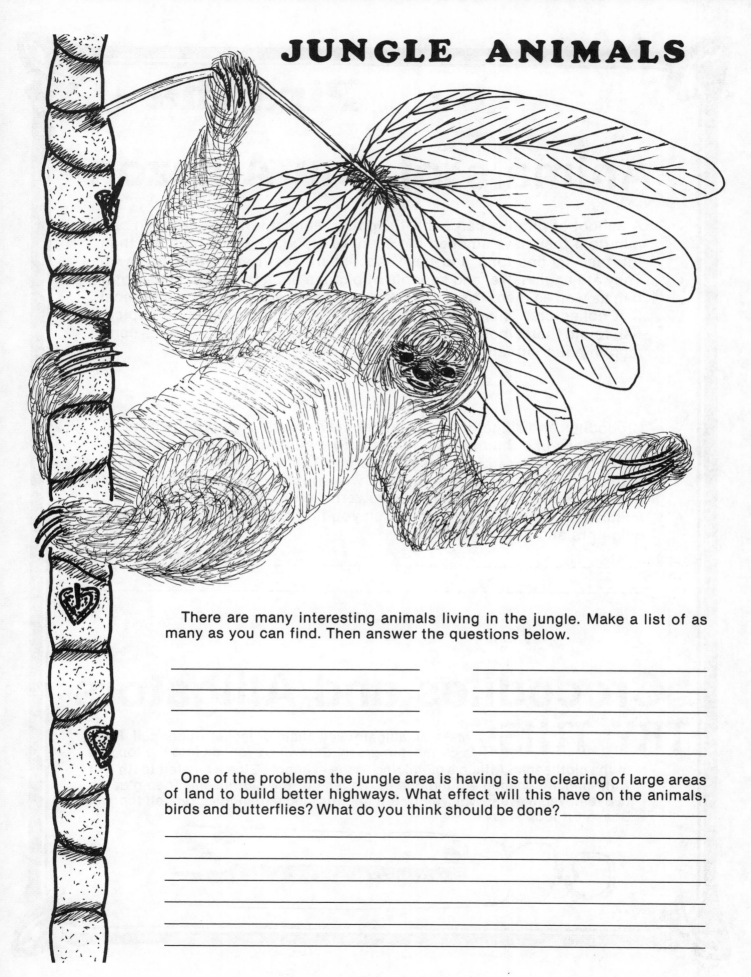

There are many interesting animals living in the jungle. Make a list of as many as you can find. Then answer the questions below.

_____   _____
_____   _____
_____   _____
_____   _____
_____   _____

One of the problems the jungle area is having is the clearing of large areas of land to build better highways. What effect will this have on the animals, birds and butterflies? What do you think should be done?_____
_____
_____
_____
_____
_____
_____

Piranha

Animals of the Amazon

A school of piranha can eat a 100-pound animal to its skeleton in one minute if they have smelled the scent of blood. They are no more than fourteen inches long and have razor-sharp teeth and powerful jaws. They are aggressive, and schools of piranha are a threat to other larger fish as well as man. The waters of the Amazon are searched by tropical fish collectors. There are more than 2,000 different kinds of tropical fish found there, including the angelfish and tetras. These fish are common in aquarium shops in the United States.

Crocodiles and alligators are found in the rivers throughout the tropical areas. They use their strong, flat tails to push themselves through the water. Their webbed feet are for steering. The sharp teeth are good for catching their prey. When old teeth wear out, alligators can grow new ones to replace them. They are powerful enough to kill and feed on animals up to the size of a cow. Can you find two differences between alligators and crocodiles?

_____

_____

Crocodiles and Alligators
TRY THIS: Make an alligator key chain or jacket zipper-pull. Paint a clothespin green. Screw an eyehook into the top end of the clothespin. Attach a metal ring to the eyehook. Glue white felt teeth (cut with pinking shears) around one of the legs of the clothespin. Then glue on two removable eyes; oval, yellow felt nostrils; and a triangular, red felt tongue. Doesn't it look scary?

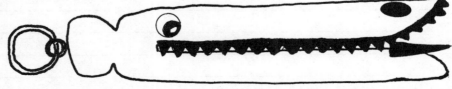

The birds of the jungle are prized all over the world for their beauty, their feathers and as pets. Often these birds are smuggled out of Brazil because it is illegal to sell them in other countries.

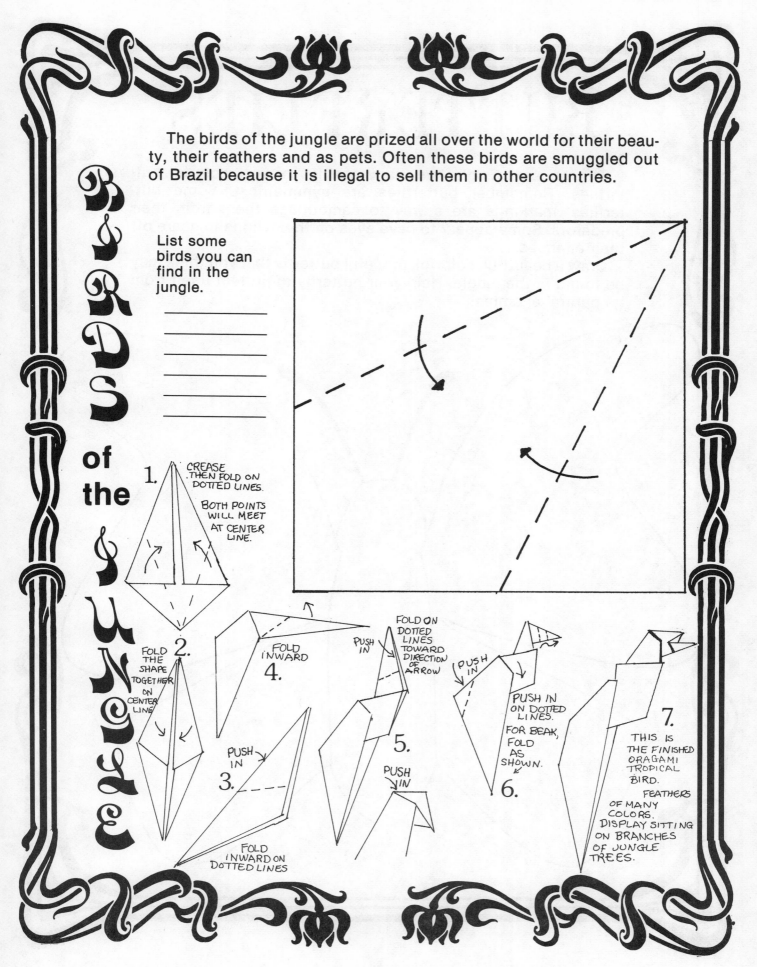

List some birds you can find in the jungle.

_____
_____
_____
_____
_____

1. CREASE. THEN FOLD ON DOTTED LINES.

BOTH POINTS WILL MEET AT CENTER LINE.

2. FOLD THE SHAPE TOGETHER ON CENTER LINE.

3. FOLD INWARD ON DOTTED LINES

PUSH IN

4. FOLD INWARD

PUSH IN

5. FOLD ON DOTTED LINES TOWARD DIRECTION OF ARROW

PUSH IN

6. PUSH IN ON DOTTED LINES. FOR BEAK, FOLD AS SHOWN.

PUSH IN

7. THIS IS THE FINISHED ORAGAMI TROPICAL BIRD.

FEATHERS OF MANY COLORS. DISPLAY SITTING ON BRANCHES OF JUNGLE TREES.

BUTTERFLIES

The Amazon jungle has many species of very colorful butterflies. Remember butterflies are **symmetrical.** Some butterflies' markings are a way to camouflage them from their predators. Some appear to have eyes on their wings to scare off their enemies.

Draw a beautiful, colorful, graceful butterfly that you think may be found in the jungle. Help your butterfly to protect itself from its natural enemies!

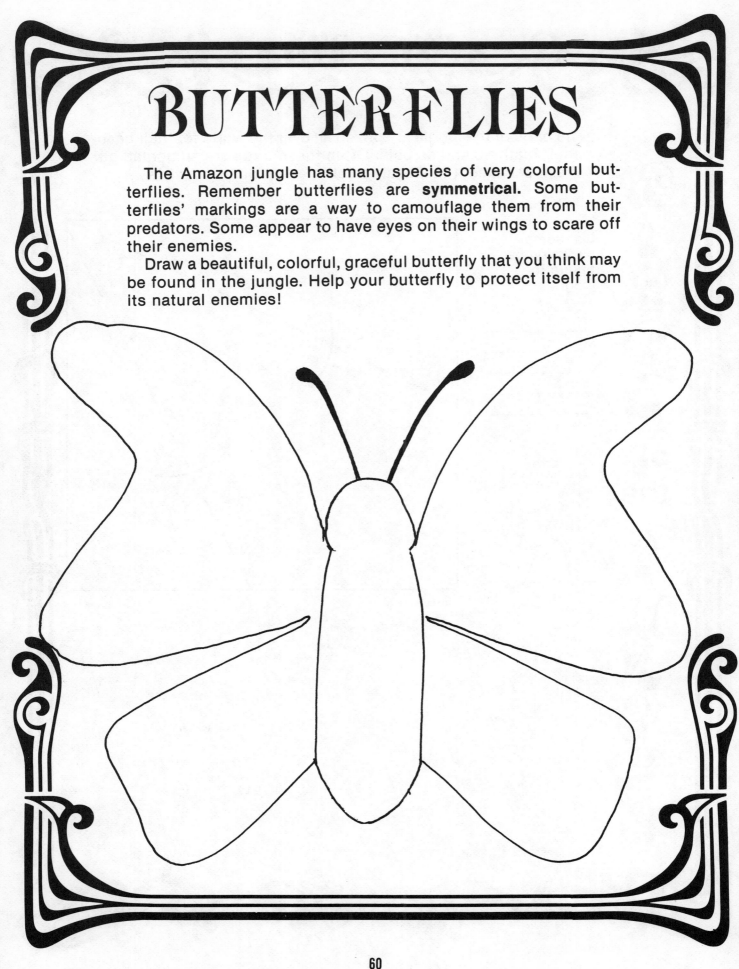

SOCCER

The national sport of Brazil is **futebol** or soccer football. Brazil has trained some of the world's best soccer players, including Pele', The King of Soccer. Pele' scored 1086 goals in sixteen years. As a player, he became the highest paid and most famous athlete in the world. He retired from soccer in 1971.

The soccer players are often asked to play for teams in other countries. Brazil has won the world soccer championship several times.

The Brazilians are eager sports fans. Huge stadiums have been built throughout the country to provide enough room for spectators.

Use the game below with a friend. You can practice some of the new Portuguese words and their meanings that you have learned, or you can practice your math facts, spelling words, etc., while playing.

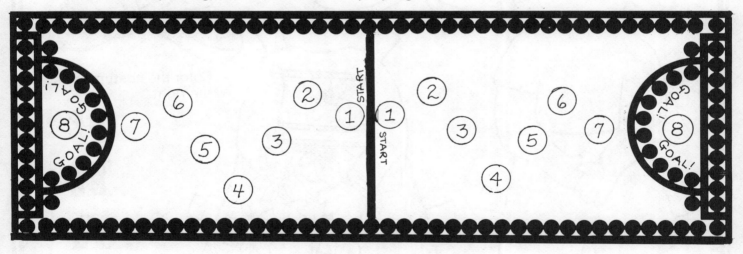

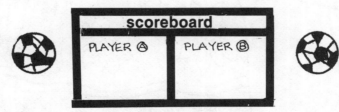

DIRECTIONS: Choose a soccer ball. Place your player on the **start** space. Whenever you give a correct answer, move your player one space. Whoever scores the most goals is the winner.

Design your own Brazilian costumes on the left. Color carefully.

Color the costumes on the right.

BRAZIL

TRAVELIN' ON...

The following is a list of additional topics that interested or highly motivated students may want to learn more about:

Manioc
Process for making chocolate (write to Hershey Foods,
Hershey, Pennsylvania for information:)
Pele'
Famous international soccer stars
Rules for soccer
Coffee production
Insects
Rio de Janeiro
São Paulo
Tapirs
Poisonous snakes
Vaqueros
Gold, diamonds, amethysts
Carnauba palm trees
Trans-Amazon highway
Flag and its symbols

TIPS FOR THE TOUR GUIDE

Klass Kickoff:

Ask a parent, another teacher, or your principal to dress in a disguise as an airplane hijacker. Have the "volunteer" enter your classroom secretly and with a hidden weapon (hand in pocket as if holding a gun or a ticking clock in a suitcase) demand the pilot (tour guide) to fly the classroom plane to an undisclosed country. The hijacker should give these directions: Fly to longitude 125 degrees North and latitude 55 degrees West. The landing strip is cut between very tall Douglas fir trees. Look for a salmon packing plant and a lumber mill. The snows will come soon and close in the airstrip. Days are very short in the fall. They need to hurry unless they want to be snowed in with the natives. What country are they going to?

Tips:

British Columbia is Canada's western most province. It is bordered by the Pacific Ocean on the west and the province of Alberta on the east. To the north is the Yukon Territory and to the south are the states of Washington and Idaho. Most of the population lives on the southern coastal region. The capital city of Victoria is located on the island of Vancouver. There the climate is mild year round. Even during the coldest months the temperature is above freezing. The island of Vancouver has lush green lawns and gorgeous gardens. The British Columbia streets are lined with flower boxes, flower beds and hanging flowerpots. Other areas in British Columbia can be bitterly cold, such as Dawson Creek where winter thermometers can dip to -50 degrees F. There is an overabundance of rain in the west and dry, sagebrush areas in the east.

The mountains of British Columbia are its most beautiful feature. Several mountain ranges form the area including the Western Cordillera and Rocky Mountains. The Fraser River is a major river in the west. The cold, clear waters rush wildly toward the sea. In the lower Fraser Valley, dairy farming and agriculture are carried on.

The heart of British Columbia's wealth is in its immense forests. Douglas fir, western hemlock and cedar are in abundance. Lumbering, wood products and paper-making are the most important industries. Mining for copper and lead is also growing in importance. The world's largest salmon fishing industry is in British Columbia.

Indians were the first inhabitants of British Columbia; now only 52,000 live there. Most people are of British descent. Vancouver has one of the largest Chinese communities in North America.

Captain James Cook landed on Vancouver Island in 1778. Simon Fraser founded the first trading post and later explored the river bearing his name. In 1858 gold was discovered in the riverbeds, and Victoria turned into a boom town overnight.

People who live in British Columbia are still asking themselves whether they should break away from the Canadian government. Some feel isolated and not well-represented in government. Some, in earlier years, also considered becoming part of the United States.

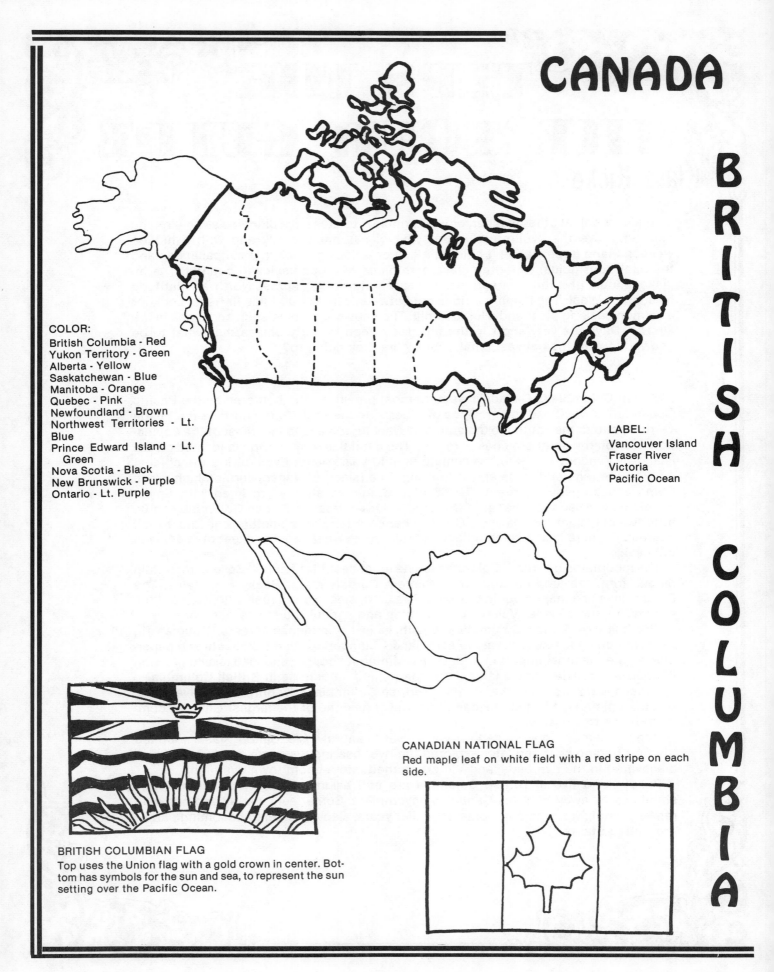

CANADA

BRITISH COLUMBIA

COLOR:
British Columbia - Red
Yukon Territory - Green
Alberta - Yellow
Saskatchewan - Blue
Manitoba - Orange
Quebec - Pink
Newfoundland - Brown
Northwest Territories - Lt. Blue
Prince Edward Island - Lt. Green
Nova Scotia - Black
New Brunswick - Purple
Ontario - Lt. Purple

LABEL:
Vancouver Island
Fraser River
Victoria
Pacific Ocean

BRITISH COLUMBIAN FLAG
Top uses the Union flag with a gold crown in center. Bottom has symbols for the sun and sea, to represent the sun setting over the Pacific Ocean.

CANADIAN NATIONAL FLAG
Red maple leaf on white field with a red stripe on each side.

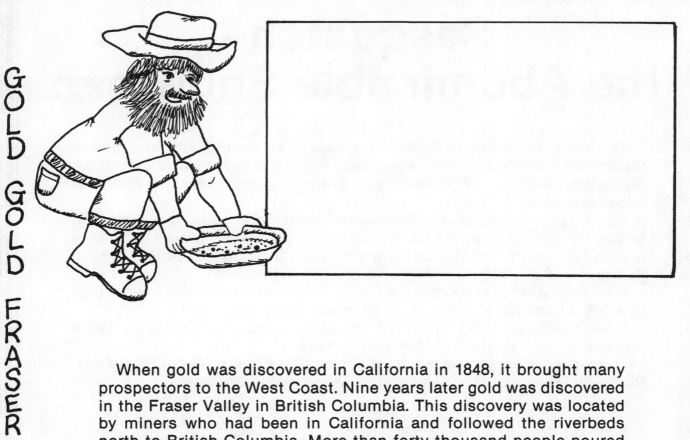

GOLD GOLD GOLD GOLD GOLD GOLD

GOLD FRASER VALLEY GOLD GOLD

When gold was discovered in California in 1848, it brought many prospectors to the West Coast. Nine years later gold was discovered in the Fraser Valley in British Columbia. This discovery was located by miners who had been in California and followed the riverbeds north to British Columbia. More than forty thousand people poured into this river area. In three years several million dollars' worth of gold was discovered. One area along Williams Creek had four thousand miners in a one-mile stretch. There was a report that one gold claim brought two thousand dollars a day, every day, for a whole year. Other miners were not so lucky. Many returned south, spirits broken. Others went into business in the area and reaped the money from the prospectors for their services or products. An account from Dawson Creek tells of a man who successfully transported a milk cow to the town and began selling the milk---for $10.00 a quart!

What kinds of services would miners want? _____

_____

What kinds of equipment would miners need? _____

_____

What kinds of buildings would Dawson Creek have in it? _____

_____

Using this information, draw a postcard above that would show a picture of Dawson Creek in the year 1858.

Sasquatch -
the Abominable Snowman

    The Indians of northern British Columbia knew about giant, wild men of the forests hundreds of years ago. These Indians called them "Sasquatch." Since the discovery of fourteen-inch footprints, the name Big Foot has become popular.

    Some people believe the Sasquatch could have come over the land bridge from Asia. It would be similar to the way wild animals may have reached North America. Since the abominable snowman is thought to have its home in Asia, Big Foot may be considered its cousin.

    It is interesting to think about the size of Sasquatch and how it would act in a city, at school, at a picnic, at an amusement park, etc.

    Draw a picture below, showing Sasquatch in one of these places. Write three sentences telling how Sasquatch feels.

    To read more about Sasquatch, research these areas: Mount St. Helens, Big Foot, Monsters, Giant Hairy Apes, Abominable Snowman.

The Endangered BALD EAGLE

The first people living in this area of Canada were the Indians. They admired eagles and considered them sacred birds that should be honored for their strength and courage. We can see how much they were admired by the number of eagles carved on totem poles. The eagle was known as the thunderbird to the Indians. They believed thunder came from the flapping of the eagle's wings. Although the eagle is not bald, its white head can give this impression when seen from a distance. It also has white tail feathers. The rest of its body is covered with dark brown feathers. The wingspan of the eagle can be six to eight feet wide. The favorite food is fish, especially salmon. It is able to see at a great height, swoop down into the water and with its powerful talons, strike, and rise with a fish. It also eats small mammals and birds. Can you guess why we say some people have "eagle eyes"? _____
_____

List the characteristics of eagles, such as powerful, strong, etc. Then try to think of an animal that may also possess these qualities. Have a campaign and try to have your animal elected as your classroom's mascot.

Characteristics	Animals

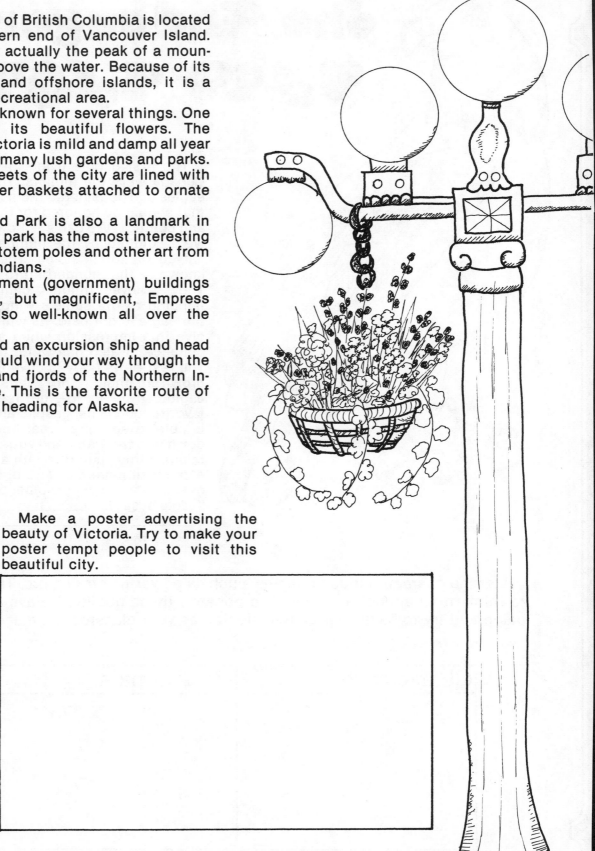

The capital of British Columbia is located at the southern end of Vancouver Island. The island is actually the peak of a mountain that is above the water. Because of its many inlets and offshore islands, it is a year-round recreational area.

Victoria is known for several things. One of these is its beautiful flowers. The climate of Victoria is mild and damp all year round. It has many lush gardens and parks. The main streets of the city are lined with hanging flower baskets attached to ornate lampposts.

Thunderbird Park is also a landmark in Victoria. This park has the most interesting collection of totem poles and other art from the coastal Indians.

The Parliament (government) buildings and the old, but magnificent, Empress Hotel are also well-known all over the world.

If you board an excursion ship and head north, you would wind your way through the mountain-island fjords of the Northern Inland Passage. This is the favorite route of many people heading for Alaska.

VICTORIA

Make a poster advertising the beauty of Victoria. Try to make your poster tempt people to visit this beautiful city.

In a World of Its Own

Carvers used cedar trees with very few knots to carve totem poles. Knots were hard to carve around. After the cedar trees were cut to the desired length, they were split in two and each pole was hollowed out to about a five-inch thickness. Then the bark was removed, and the carver was ready to begin.

Totem poles were painted with fresh salmon eggs mixed with saliva and minerals. Red paint came from the mineral hematite, black from carbon or graphite, and green from copper ore.

The tallest totem pole in the world is the Totem Pole of Sakau'wan. It is eighty feet six inches tall.

TOTEM POLES

A totem pole is made up of figures that may portray a dream, a special event, or a death. The poles were never worshipped but were respected as guardian spirits. They were carved originally with stone tools using western red cedar logs.

In central British Columbia, the Kwakiutl are still very expert totem carvers. Their poles were brightly painted and carved with much detail. Thunderbird, the ruler and master of the skies, is shown most often and usually has his wings outspread.

Totems were used to tell stories when no written language was known.

Only the greatest and wealthiest Indian chiefs had totem poles.

Make your own totem pole. Use grocery bags stuffed and stapled. Paint and add details to the stuffed bags to represent the totem symbols. Staple them together. You may need to lean this pole against a wall or secure it with wire to the ceiling to help it stand.

Another way to make a totem is to use different shaped boxes and decorate them to resemble a character on a totem. Paint them and assemble them into one large totem pole.

Empty ice-cream tubs are also easy to decorate and stack into an interesting totem pole.

Could you create a totem that would represent your school? Think of the symbols and what characteristics they had. Perhaps you could design and construct a totem for the school office.

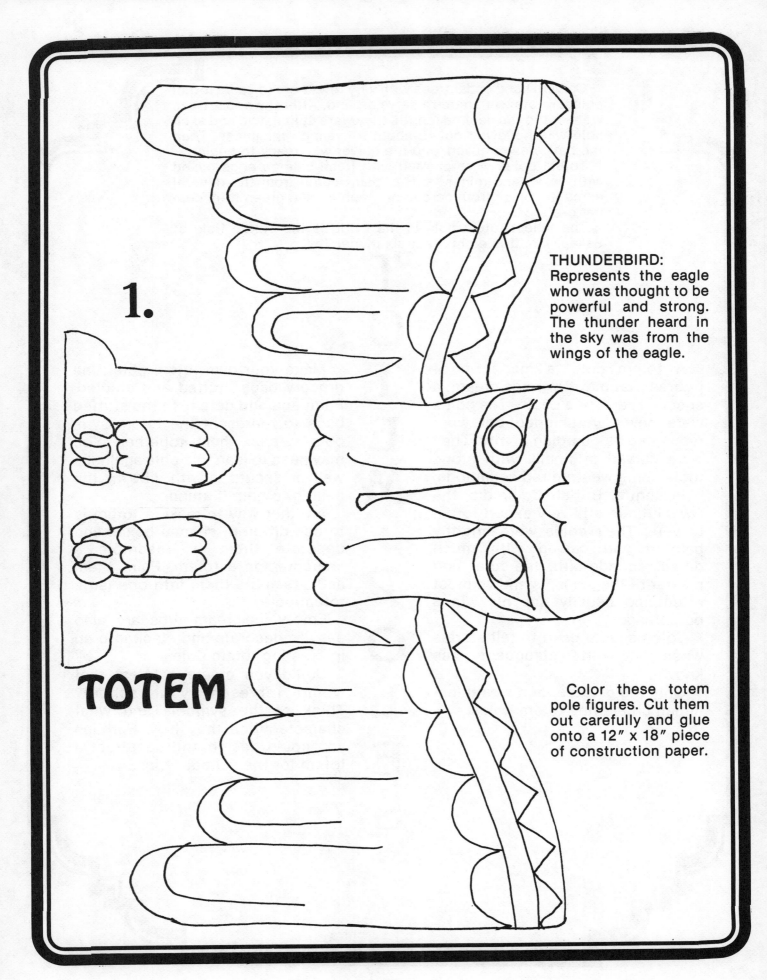

1.

TOTEM

THUNDERBIRD:
Represents the eagle who was thought to be powerful and strong. The thunder heard in the sky was from the wings of the eagle.

Color these totem pole figures. Cut them out carefully and glue onto a 12" x 18" piece of construction paper.

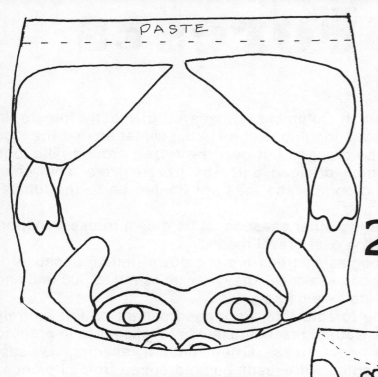

Could you write a story about this Indian Chief's life? Use the information about each of the figures to give you an idea.

2. FROG: The frog is found near water. It has large eyes and is quick to capture its victim.

POLES

3. CLANSMAN: This is one of the Indians from the clan who was swallowed by a huge halibut, and when caught the clansman was found inside its stomach.

4. SEA MONSTER: This is the fish and bird sea monster that swallowed the clansman.

Forests

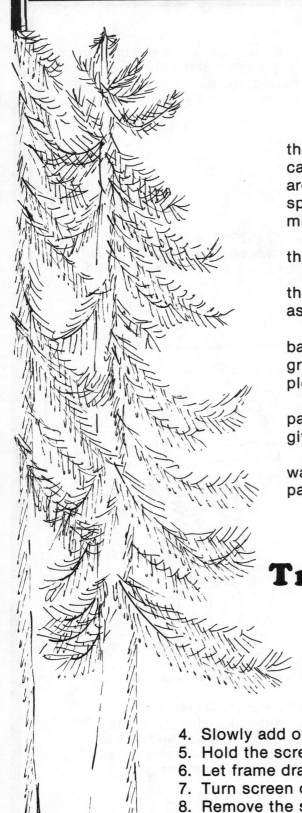

British Columbia is made up of mostly forests. In the east, logging is done in the winter so that the logs can be dragged out over the frozen ground. The logs are then dragged onto the frozen rivers, and when spring comes, the logs are floated on to the lumber mills.

During other seasons, huge diesel trucks transport the logs over gravel roads.

Douglas fir trees are cut down (felled) because of their colossal size. They may be as tall as 300 feet and as wide as a garage.

The forests that are logged cannot be left entirely bare. Natural reseeding needs to take place in order to grow new trees. Often natural seeding is supplemented with seeds being dropped from airplanes.

Much of the wood and sawdust is used for making paper. A simple method for making your own paper is given below.

On the next page is a paragraph frame. You may want to write your finished copy on your self-made paper!

Try this:

1. Make two small screened wooden frames (about 6″ x 6″).
2. Tear colored (or white) facial tissue into small pieces.
3. Add warm water and mix with mixer or blender.
4. Slowly add one cup of dry starch.
5. Hold the screen frame and dip into the water-paper mixture.
6. Let frame drain.
7. Turn screen over onto a paper blotter.
8. Remove the screen and put another blotter on top.
9. Iron the paper dry. Remove the blotters.
10. Use your new paper to write your best copy of your paragraph frame.

# of Fir

Fill in the blanks with your own words. These words should answer the questions given below the blanks. When you are finished, check for spelling errors. Rewrite your story and read it to the rest of the class.

British Columbia is made up mostly of forests of
_____trees. They are _____.
(what kind of?)   (what?)
Lumbering is important because_____. The
                                    (why?)
lumber companies replace the trees they have cut down by
_____. Some products we get from trees are
    (how?)
_____.
    (what?)

List all of the products we get from paper on the lines below. Would you say we use a lot of paper?
_____

1._____
2._____
3._____
4._____
5._____
6._____
7._____
8._____
9._____
10._____
11._____
12._____

FISHING

Kinds of Nets

Gill nets float on the surface of the water and hang down like a curtain. The salmon swim into this curtain, their heads are caught in the mesh and their bodies are too big to go through. When they try to back up, their gills are caught.

In the rivers that flow to the oceans, spring brings the salmon to **spawn** or lay their eggs. After they have spawned, the salmon die. Scientists believe the salmon use their sense of smell to find their way back to the exact rivers where they were hatched. In British Columbia, salmon are the most important fish caught. They are taken out of the rivers with gill nets, seines, or trolls.

Gill Net

The seine nets float, too. They are laid out in a circle. As the fish enter the circle, the net is gathered up under them. The fishing crew scoops the fish out of this net "basket" with scoops called **trailers.**

Open Seine Net   Closed Seine Net

A troll is a long line with hooks attached. It is dragged behind a boat moving slowly through the water.

Trolling

FOR SALMON

Most of the salmon caught are canned or frozen. The heads and tails are used for fertilizer. Every part of the fish is used.

There are several manmade structures to help the salmon return to their birthplace to spawn. One of these structures is called a **ladder.** Find out more about it. Can you tell how it works?

How do you eat salmon? Does your family have a special recipe? Why do canners include the bones of the salmon? Below is a native recipe for preparing ocean salmon.

Dried Salmon

Split salmon, cut off head, and remove backbone. Cut gashes in red part of salmon leaving skin whole. Wash, hang on racks to dry outside. Sometimes the salmon is partly dried outside, then taken in, dipped in flour, put into a pan and fried in whale oil.

Are there other fish that also return to their birthplaces to die? You may want to make a chart of different kinds of ocean fish found on the shores of British Columbia.

The life of the salmon is interesting. Make a mural showing the life cycle of this fish.

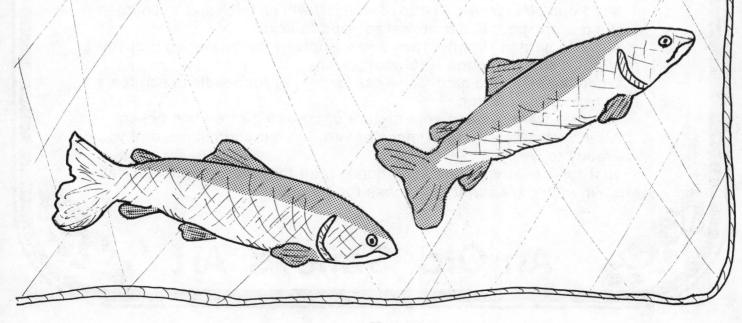

SCRIMSHAW

Scrimshaw is a dying art among the natives of British Columbia. It is the art of scratching pictures into shark's teeth or whale bone. Often the ivory tusks of the walrus were also used. The sailors were known for this art form.

Today, antique scrimshaw is highly priced and regarded as a unique art form.

You may want to try scratching your scrimshaw in plaster shapes rather than whale bone!

Try this:

Mix 1 cup plaster with ⅔ cup of water. It will be thick and will harden fast, so drop spoonfuls onto waxed paper quickly.

Allow to harden slightly. Then make a hole in the plaster so that you may put a string through it to wear.

When the plaster is hard, use a nail, paper clip, or needle to scratch a picture into the plaster.

You may want to paint your picture or try dark paste shoe polish.

Attach a leather string or piece of yarn to your scrimshaw, and you are ready to wear it!

Just for fun—try boiling soup bones until they are free of meat, fat, etc. When dry, try scratching these for scrimshaw.

An Old Sailor's Art

IN SEARCH OF WILDLIFE

```
B A C D G I  J M O B O B C A T E B H K O
R I R S U M U S K R A T V K D W L X A Z
O A S B I  J C G E E S E T L E D A E T F
W G H O B I  G H O R N S H E E P C K A L
N P M X N Y L E N I  O P S Q R S K T O U P
B V O W X Z S A P B M O O S E S B C G D
E F E R F I  G U H I  O J P K L R E M N P
A O Q S O U C T R G Q U A I  L E A V I  W
R X A P Y R B Z N C M I  N K D V R E A F
G H R I  O J L A T R O U T M K A S N T O
P O R P Q K I  L L E R W H A L E S S N D
P T G O L D E N E A G L E S A B U W U A
S E L G A E D L A B B D R L G E H C O F
W A C N X S E A G U L L S Y J W K Z M I
U S A L M O N S Y T P N Q L O S M K R O
A C A R I  B O U R A E B Y L Z Z I  R G W
K B K E Z J C O F I  Y D F N H L X P M G
```

Locate the wildlife
in the word search.

sea gulls
elk
moose
caribou
mountain goat
bighorn sheep
muskrat
beaver

black bear	quail	killer whale
brown bear	bison	porpoise
fox	lynx	deer
golden eagle	mink	seal
grizzly bear	porcupine	bobcat
bald eagle	Canadian goose	wolf
duck	trout	otter
geese	salmon	panther

British Columbia

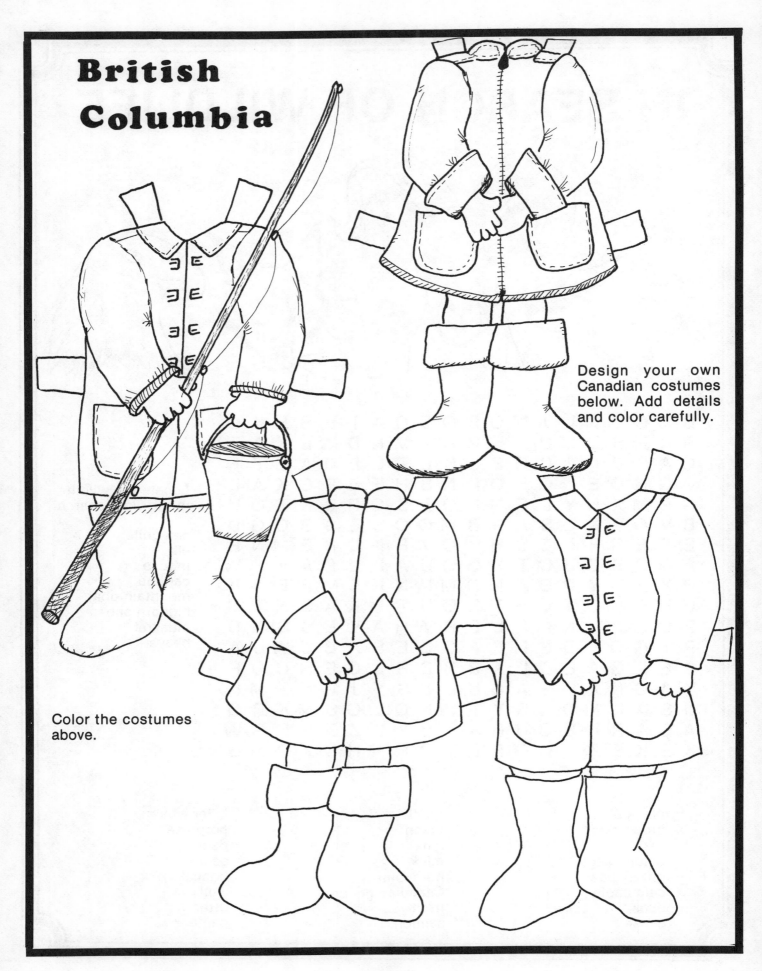

Design your own Canadian costumes below. Add details and color carefully.

Color the costumes above.

TRAVELIN' ON...

The following is a list of additional topics that interested or highly motivated students may want to learn more about:

Captain James Cook
James Fraser
The Chinese in B.C.
People of Cariboo and Kootenay
Dawson Creek
Japanese current
Robson Falls
Salmon fishing
Lead mining
Kinds of evergreens
Tubing on the Fraser River
Totem poles
The importance of the railroads
Hudson's Bay Company
Inuit (Indians or Eskimos)
Flag of Canada
Sports
Canada and England's relationship
National parks in British Columbia

TIPS FOR THE TOUR GUIDE

Klass Kickoff:

Use the classroom doorway or a refrigerator box with opposite sides cut out to make a walk-through metal detector similar to one at an airport. Each person passing through must give one piece of information about Germany. (For example, spell Germany, name the capital, famous car manufacturer, etc.) You may want the students to have a homework assignment to find this information the night before you begin your unit of study. This also involves the parents, and you may get several parent volunteers who have visited or lived in Germany that will talk to your class.

Tips:

West Germany's official name is the Federal Republic of Germany. Bonn is the capital of the country. Not only is Bonn the capital of Germany but also well-known as the birthplace of composer Ludwig van Beethoven. The city of Oberammergau has kept a 300-year promise to perform the Life of Christ every ten years if the Black Plague ended. More than three-fourths of the population lives in cities which are connected to the countryside by the autobahn (freeway). A majority of the sixty million people are Protestant and speak the German language. The Bavarian Alps include the beautiful Black Forest scenery. The Rhine River forms its boundary. One of the favorite sports is downhill skiing. The Bavarian Alps provide not only skiing slopes but beautiful, breathtaking scenery. There are many castles on the hilltops and vineyards along the Rhine. This is the land of thatched roof houses, cuckoo clocks and Mercedes-Benz automobiles.

The Ruhr Valley is the heart of the farming area. The main crops grown are rye, wheat, barley, oats, potatoes, and sugar beets. Livestock and fishing are also important to the German economy as is the chemical industry. Pharmaceuticals, synthetic fibers, fertilizers, and industrial chemicals are manufactured. Fine tools, cutlery and precision instruments, such as cameras, clocks and watches are important trade items. The city of Nuremberg is famous for its Toy Trade Fair. Berlin is divided between East and West Germany. The city of West Berlin has many old people who lived there before World War II. Young people leave because the city is surrounded by East Germany and really isolated from the rest of the country.

The children often look like the children in Grimm's fairy tale, "Hansel and Gretel." The girls wear **dirndl** dresses which are bright print jumpers with lacy blouses. The boys wear **lederhosen** or leather shorts with suspenders. Older children wear jeans and T-shirts. They attend **gymnasium** (high school) or work as apprentices.

The German people enjoy singing, dancing, and eating. The currency used is the **mark** which is equal to approximately one U.S. dollar. The average per capita income is $9,278 per year.

EUROPE

Label the European countries:

West Germany
East Germany
Netherlands
Belgium
Luxembourg
France
Spain
Portugal
Italy
Switzerland
Austria
Hungary
Czechoslovakia
Poland
Sweden
Norway
Denmark
United Kingdom
Ireland

Using the scale of miles, measure 300 miles in all directions from *Bonn, West Germany. Use different colored dotted lines to show your flight plans.
Will you always land on terra firma? _____

N
nw   ne
W   E
sw   se
S

scale of miles
0   100   200   300   400   500

BEETHOVEN

Bonn is the city where the famous composer, Ludwig van Beethoven was born in 1770. His birthplace is now a famous museum. A music festival is held every two years in honor of him. All 140,000 residents of Bonn appear to be music lovers because the concert halls are always packed with people.

Beethoven was deaf when he composed his greatest works. He conducted the famous **Ninth Symphony** himself without ever hearing it. The first four notes in his **Fifth Symphony** were used as a radio signal during World War II.

BELLS

In the mountain areas, a favorite type of folk singing includes yodeling. This is a kind of warbling done with one's voice. It takes much practice to perfect yodeling.

Try saying these words slowly at first and then speeding up. Perhaps you can become a yodeler, too!

"Hol-di-ri-a-du-i-a-i-ri-a-ho,
Hol-di-ri-a-du-i-a-i-ri-a-ho!"

This is part of the chorus to the "Swiss Yodel Song" written by Wiedmer in 1840. You may want to learn the verses, too.

BELLOWS'

In the mountain areas of Bavaria, you can see the cows with bells hung around their necks. These colorful painted bells help keep the cattle together and also announce the return of the cattle and their tenders from the grazing lands.

Huge cowbells and very long horns are used during celebrations. The German people enjoy music and need little encouragement to perform! Can you write some new German verses for the tune, "Three Blind Mice"?

Fishing at the sea. (repeat)
The Alps are higher than me.
  (repeat)
(three lines of your own...)
Fishing at the sea.
The Alps are higher than me!

BIRTHDAYS

Hundreds of years ago, the people of Germany started having a yearly party with cake, presents, and all the trimmings. The **kinderfest,** or children's party, soon spread to other European countries and later, as these Europeans settled in the United States, the custom also immigrated.

The German people believed that the candles on the birthday cake had magic powers to make wishes come true. This is the reason we try to blow out **all** of the candles!

Sketch your favorite birthday cake and put on the "magic" candles.

Make Your Own Black Forest Cuckoo Clock

Trace the clock pattern on the next page and duplicate it for this math/art project. Cut a six-inch square pattern from railroad board.

Glue this cuckoo clock pattern onto the railroad board frame.

Add a yellow pompon with an orange construction paper beak and black eyes for the cuckoo. Two orange pipe cleaner pieces will work nicely for the legs.

Make the clock hands from aluminum foil or gold foil wrapping paper. Attack with a brass fastener so that the hands can be turned.

If real pinecones are available, tie yarn on the cones and attach to the bottom of the clock. Otherwise use brown construction paper cones.

The cuckoo clocks can now be used to practice telling time or for Roman numeral recognition.

THE

BLACK FOREST CUCKOO

CUCKOO CLOCKS

Fill in the blanks:

I = ____   II + V = ____

II = ____   XI + III = ____

III = ____   IX + I = ____

IV = ____   VII + III = ____

V = ____   VI + II = ____

VII = ____   IV + II = ____

IX = ____   X + III = ____

X = ____   XII + IV = ____

XII = ____   IX + VI = ____

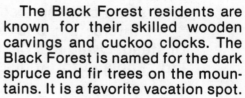

The Black Forest residents are known for their skilled wooden carvings and cuckoo clocks. The Black Forest is named for the dark spruce and fir trees on the mountains. It is a favorite vacation spot.

Most cuckoo clocks chime and sing on the hour. You will also notice the clock face uses Roman numerals rather than the familiar Arabic numerals.

87

CASTLES

There are many castles in West Germany. Some of the most famous are the Heidelberg Castle, Wilhelmstal Castle, and the Sans Souci Castle. A castle was proof of the owner's power and rank. If the owner was very rich, he probably had more than one castle in different parts of the country. The owner had to keep up with the latest warfare equipment in order to protect his castle(s).

His family lived in the castle along with a hundred or more servants and officials.

The Great Hall of the castle was the largest and most important room. There was not much furniture. The owner held court there and the people of the household ate and celebrated there. The most important servant in the castle was the steward. He watched over all affairs and saw that each official carried out his job.

An interesting and easy-to-read book about castles is by R.J. Unstead, **Living in a Castle,** c. 1971. It was published by A + C Black, Ltd., England for Addison-Wesley, U.S.A.

Try to find out more about the famous castles of Germany. Could you use this information to help you build your confectionery castle on the next page?

CASTLES OF CONFECTIONERY

Make a confectionery castle. This castle would be very similar to a gingerbread house. You can use quart-sized milk cartons, cut to various heights, for the foundation of your castle.

Cover the cartons with graham crackers. Use the icing recipe to "glue" everything together. Place the icing in small paper cups with plastic knives, so each child can "glue" his own castle. Then decorate with candies. Some suggested candies are green leaf gumdrops, red licorice twists, red licorice strings, red cinnamon hots, lemon drops, pillow mints in assorted colors, and other assorted jimmies.

Icing Recipe
(Makes 2½ cups)

3 egg whites
½ teaspoon cream of tartar
1 package (1 pound) sifted confectioner's powdered sugar

Beat the egg whites and cream of tartar until frothy and foamy. Gradually add sugar, while continuing to beat until the icing stands in firm peaks and is still enough to hold a sharp line when cut through with a knife.

NOTE: Frosting dries out quickly. Keep bowl covered with several moist paper towels, working with small amounts at a time. The frosting may be refrigerated for several hours if frosting is interrupted or you want to make it ahead of time.

The **Grimm's Fairy Tales** were written by Jacob and Wilhelm Grimm. These stories were old folk tales told for many, many years by grandmothers to their grandchildren. The Grimm brothers wrote down the stories and published their first collection in 1812. Some of the most popular fairy tales told to the Grimm brothers were "Cinderella," "Rumpelstiltskin,""Hansel and Gretel" and "Snow White and the Seven Dwarfs."

In each of these fairy tales there are several things in common. Fill in the information in the chart below. Do you need to read the stories again to refresh your memory?

Then fill in some details of your own in the last column. You will use this information to write your own fairy tale.

	Cinderella	Rumpel-stiltskin	Hansel and Gretel	Snow White	Your Story
Magic Power					
Talking objects or animals					
A task to be done					
Use of numbers 3 or 7					
Wish or dream comes true					

90

Now write your own fairy tale using the information you marked on the fairy tale chart on the opposite page. Be sure to give your characters names. Perhaps you will have room for an illustration at the bottom of this page.

FANTASY

Try this:

Use the letters in your name and make them into things that have special meanings to you.

OCTOBERFEST

The greatest German festivities take place at the end of September. It is known as Octoberfest! This celebration goes on for sixteen days. Amusement parks, yodelers, huge tents, bands, folk dancers, mountains of food, rivers of beer are all features of the Bavarian Octoberfest!

Plan an amusement park for the German Octoberfest. What kind of rides would you have? What kind of musical entertainment? What about food and drinks?

Draw a map of your Octoberfest Amusement Park on a 12″ x 18″ piece of construction paper. Cut out the map key and compass "stein" from this page and glue it to the construction paper.

What would be a fun and catchy name for your park? Some suggested amusements might be:
 Black Forest Log Ride
 Matterhorn Roller Coaster
 Cuckoo Clock Scrambler
 Hansel and Gretel Stage Show
 Sauerbrauten and Dumplings Food Cart

WORLD WAR II

After World War I the German people were often poor and hungry. Men searched for jobs and food. During this time, Adolf Hitler promised the people prosperity. Hitler was a member of the Nazi party and became the dictator of Germany. His word became law. Anyone who disagreed with him was put in prison or killed. Hitler wanted to rule the world. World War II began when Germany invaded Austria, Poland, France, and England. Find these countries on a map.

At first the German armies could not be beaten. Then the countries of Great Britain and Russia entered the war. The attack by the Japanese on Pearl Harbor brought the United States into the war. These countries and all of their allies (friends) proved to be stronger than Hitler's armies. Germany was defeated.

The great German Empire was divided into East Germany (controlled by Russia) and West Germany (a democratic government). These two separate countries remain today.

Much more information is available on World War II. If you are interested, find out more about it.

Below is a game to be played by two people. Roll a die to see who starts first and the number of spaces to move. Object of the game is to follow the tank tracks to reach the Dynamite Victory first. **Viel Gluck!**

Color the costumes at the top.

GERMANY

Design your own German costumes at the right. Add details and color carefully.

TRAVELIN' ON...

The following is a list of additional topics that interested or highly motivated students may want to learn more about:

World War II
Volkswagen car manufacturing
German toys
School system
Christmas in Germany
Skiing
Matterhorn
German language
Foods
Mountain altitude effect on crops
Woodcarving
Edelweiss flower
Ancient castles
Transportation
Easter celebrations
Read unfamiliar fairy tales
History of the hymn "Silent Night"
Gingerbread
Iron Curtain
Dachshund dogs bred in Gergweis
Echoes
Johann Sebastian Bach
Johannes Brahms
Ludwig van Beethoven
Richard Wagner
Martin Luther
Johann Gutenberg
German inventions

EGYPT

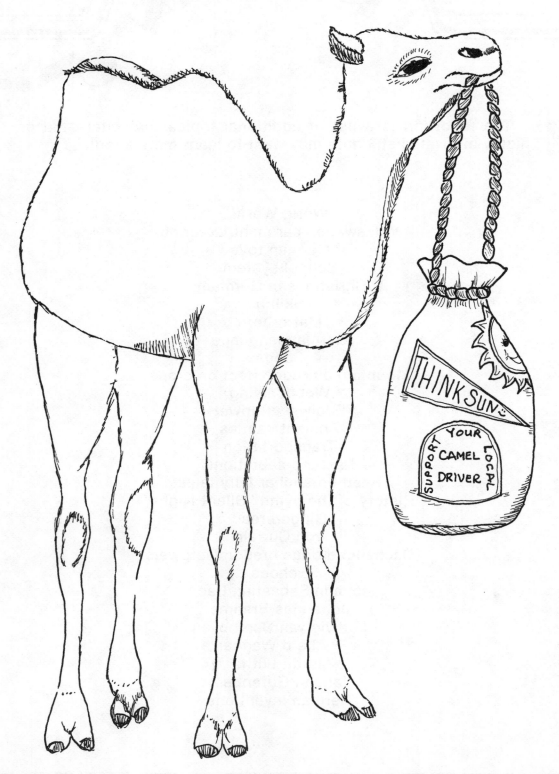

THINK SUN

SUPPORT YOUR LOCAL CAMEL DRIVER

TIPS
for the TOUR GUIDE

Klass Kickoff:

"Unearthing the Tomb of Pharaoh Goldah Kandee." Divide your class into groups of approximately five to six students. Enlarge the map below or use a map of your classroom. Hide a "treasure" in the room. The treasure could be gold-covered candy coins in a pyramid box or a box with pyramids on it. Count off the number of feet from each X on the treasure map and write this information on the map. You may want each group to go separately and time how long it takes the group to discover the treasure. Don't consume the "Goldah Kandee" until **all** groups have had a chance to find it. You could also hide a treasure for each of your groups and have each treasure map locate that treasure in a different place in your classroom. Using the latter idea, all students could work at the same time. Remind the students that they are archaeologists who work carefully and accurately. Warn them that there may be a curse that goes with the treasure! They may be taking their lives in their own hands!

Tips:

The history of Egypt is interesting and inspiring when you consider the many discoveries and inventions from early times.

The official name for Egypt is the Arab Republic of Egypt. This name was derived after a new constitution was signed in 1971. Before this time, the area was known as the United Arab Republic. The land in most of Egypt (96%) is desert, leaving only 4% for the survival of the approximately forty million people. Egypt has not had the good fortune of finding large oil deposits on its shores; and, therefore, the economic standard has not increased at as rapid a rate as in neighboring counties. The currency used is the **pound** and the income of the workers is not high—about $170 per year. The most important crop in Egypt is cotton. The mild climate and irrigation help to grow about three crops of cotton per year.

The Nile River plays such an important part in the lives of the Egyptians. Most of the fertile land and, consequently, the population are found on the east and west sides of the Nile River.

Cairo, the capital of Egypt, is also the largest city in Africa and the busiest. Its seaports are important to the economy of the country. There are many jobs available to natives in this large city. There is much overcrowding, also. It is the religious and cultural center of Egypt.

The Bedouin nomads are the sheepherders of the desert. They also roam the area around the Sinai Peninsula. This area seems to be at war at all times. The Suez Canal (100 miles long) is also important to the entire world as it connects the Mediterranean and Red Seas. It saves ocean-going vessels thousands of miles by not having to go around the southern tip of Africa.

The country of Egypt is the size of the states of Texas and New Mexico combined. The deserts cover 96% of the country. This means that if you could divide Egypt into 100 equal pieces, ninety-six of them would be desert and four would be green and wet.

Be sure to locate Egypt on a map of Africa and on a world map.

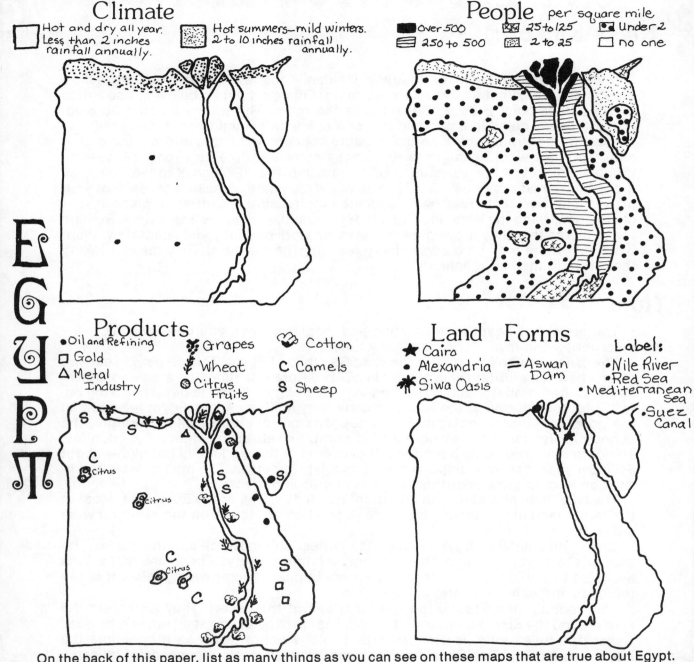

Climate

☐ Hot and dry all year.
Less than 2 inches
rainfall annually.

▨ Hot summers—mild winters.
2 to 10 inches rainfall
annually.

People   per square mile

■ Over 500   ▨ 25 to 125   ● Under 2
▤ 250 to 500   ▨ 2 to 25   ☐ no one

Products
● Oil and Refining   🌿 Grapes   🤝 Cotton
☐ Gold   🌱 Wheat   C Camels
△ Metal   ◉ Citrus   S Sheep
   Industry   Fruits

Land Forms
★ Cairo   = Aswan   Label:
● Alexandria   Dam   ● Nile River
🌴 Siwa Oasis   ● Red Sea
                              ● Mediterranean
                                 Sea
                              ● Suez
                                 Canal

On the back of this paper, list as many things as you can see on these maps that are true about Egypt. Then answer the questions below.

1. What area of Egypt has the most rainfall? (N,S,E,W)_____
2. What area of Egypt has the most population? (N,S,E,W)_____
3. Where is most of the food grown to feed the people?_____
4. Where would you go to see the oil industry?_____
5. Locate the Nile River. Why is the Nile so very important to the people? List four ways on the back of this paper.

hieroglyphics

The Egyptians gave the world writing in the form of hieroglyphics. The tombs of kings and queens were decorated with scenes of daily life and described by hieroglyphics.

In later years people had little interest in deciphering hieroglyphics because they did not think it was actual writing but merely a secret code used by Egyptian priests.

In 1799 some of Napoleon's soldiers discovered a stone tablet near the Rosetta mouth of the Nile River. They named the tablet the Rosetta stone. On the tablet was information written in three languages: Egyptian hieroglyphics, Egyptian demotic, and Greek. The Greek language was still well-known, so scholars could translate the hieroglyphics.

There are several hieroglyphic symbols on the edge of this paper. Can you guess what any of them mean? _____

Now use your symbols to send a message to one of your friends. You may write it below. Remember, you do not need a symbol for **every** word. One symbol could represent a complete thought.

Make up a symbol for these words:

bird		man		camel	
home		talked		ran	
tent		king		gold	

Now use your symbols to send a message to one of your friends. You may write it below. Remember, you do not need a symbol for **every** word. One symbol could represent a complete thought.

99

PYRAMID OF POWER

One of the Seven Wonders of the World is the Egyptian pyramids, particularly the Pyramids of Giza, located near Cairo.

It took over two million separate blocks of limestone and over twenty years of labor to build this pyramid. It is 768 feet square. It was built about 4500 years ago. It faces the points of the compass: north, south, east, west.

If you were to visit the Giza Pyramid, you could climb it with a guide.

Remember these pyramids were the sacred burial chambers of the pharoahs, who were worshipped like gods. There could be many valuable and interesting things inside the pyramids.

Scientists and archaeologists are still baffled as to how the huge limestone blocks could have been transported from the quarries and then hoisted into the shape of the pyramid.

Try this:

Try to design a machine that could help the laborers move and lift these huge limestone blocks. Remember there is no electricity, gasoline engines, conveyor belts, etc. The Egyptians were clever and imaginative people!

The sphinxes are other "wonders" of Egypt. They are creatures appearing to be half-human and half-lion. The Great Sphinx of Giza is a monument about 240 feet long. Why did the Egyptians build them? No one knows for sure. Their name means "lord" or "ruler." Perhaps they were used to guard the entrance ways to the pyramids and temples. It is thought that the Great Sphinx of Giza may have been built to guard the entrance to the Valley of the Nile.

Try this:

Draw a creature that is half-human and half-animal to guard the entrance to your school. Think of the qualities the creature should have in order to guard a school. (strong, smart, sense of humor, etc.)

SPHINX OF SAFETY

The Tomb and Treasures

During the early years of Egypt, the kings, queens, and other important people were buried in tombs when they died. The size and magnificence of the tomb showed how powerful and loved the ruler was by his people. It was believed by the Egyptians that when someone died he/she would come to life later in another world. Because of this belief, people were buried with many of their belongings and riches. The bodies were mummified. This was a special process of preserving the bodies, some of which lasted hundreds of years.

Tutankhamon (toot′ angk . a′mon) was king for only nine or ten years and was crowned when he was only nine years old. When reading hieroglyphics, scientists have found out that King Tut was quite tall (5′ 6″) for an Egyptian. He was married when very young. His wife was chosen by his family. He died when he was only eighteen years old and no one knows for sure the cause of his death.

Often the tombs of the kings and pharoahs were built in the area of Egypt known as the Valley of Kings. Most of the tombs had been robbed and the art treasures stolen and sold to art dealers many, many years ago.

Howard Carter was an archaeologist who was determined to find the tomb of the boy-king Tutankhamon (Tut, for short). In 1922 Carter did find the entrance to Tut's tomb, but was very sure the tomb had been robbed many years before. He was right about finding the tomb, but wrong about it having been robbed. It was still filled with riches: gold chariots, gold beds, jeweled weapons, alabaster bowls, battle trumpets, etc. It was a breathtaking and overwhelming discovery!

Today many people view King Tut's treasures and are still reminded of the beauty and respect shown to this very young and dearly loved king.

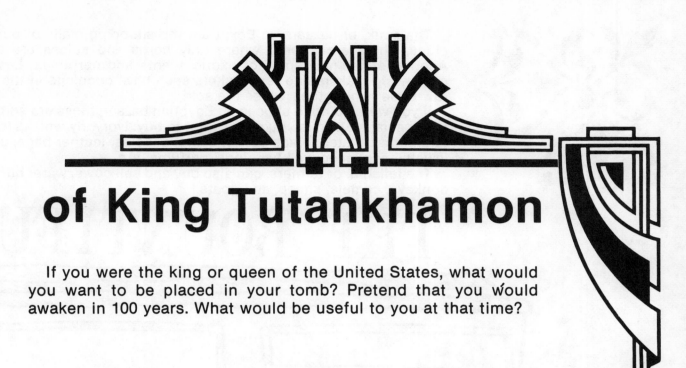

# of King Tutankhamon

If you were the king or queen of the United States, what would you want to be placed in your tomb? Pretend that you would awaken in 100 years. What would be useful to you at that time?

Make a list below of the possessions you would want, and tell why they would be important to you.

Possession

Why is it important ?

1. _____ _____

2. _____ _____

3. _____ _____

4. _____ _____

5. _____ _____

6. _____ _____

7. _____ _____

8. _____ _____

9. _____ _____

10. _____ _____

The **souk,** or bazaars, in Egypt are the shopping malls of our big cities. There are streets where only herbs and spices are sold. Another street may smell of exotic scents and perfumes. Leather goods, rug makers and tent makers each have openings in the narrow street where their wares are sold.

If you were going to shop in an Egyptian bazaar, these are some of the special souvenirs you might want to buy: ivory, fly whisks (swatters), brass plates, brass bowls, camel saddles, leather bags, silver jewelry, appliqued wall hangings, perfume, etc.

The **fellahin,** or farmers, can also buy and sell cows, water buffalo, donkeys, camels, sheep, and goats.

THE BOUNTIFUL,

Water and soft drinks are sold by merchants who sing or shout their wares. The market is a noisy place as the shoppers and merchants haggle and bargain (**fissal**) at the top of their lungs.

The market is not just a place to buy and sell; it is an important social event. Friends and relatives meet and exchange news and gossip. Everyone dresses in his/her finest clothes.

Also in the marketplaces, dancing girls dance to the music of **kanouns,** a stringed instrument, and snake charmers perform for money tossed to them by passersby.

Make a sign above each of the shops. Display what is being sold in the doorway. The shopkeeper may need to be near the shop trying to talk shoppers into buying his merchandise.

SNAPPY SANDALS

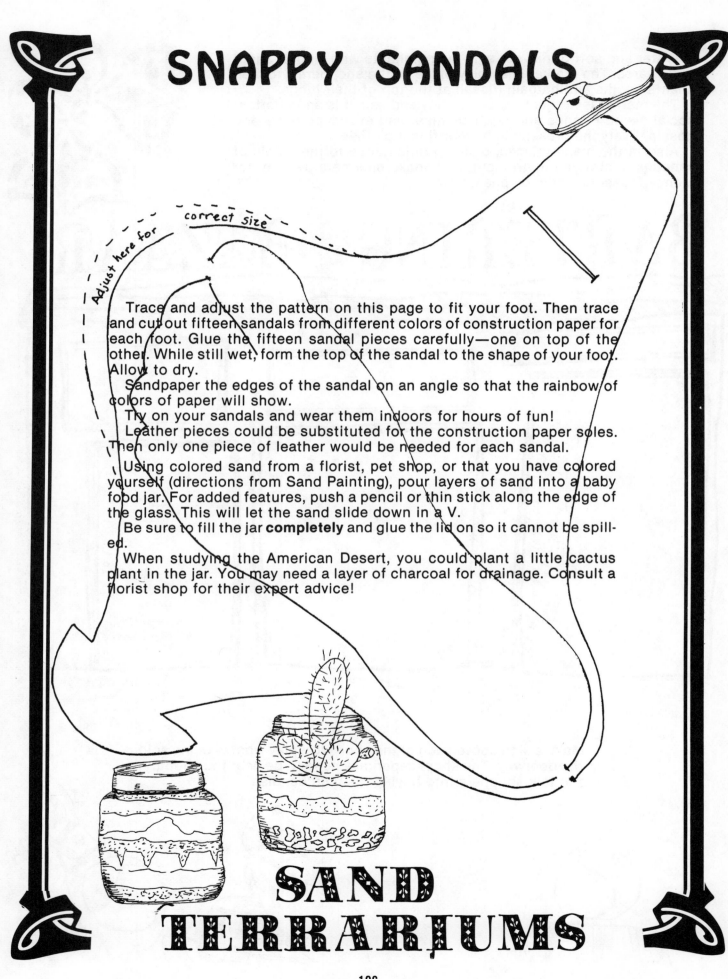

Trace and adjust the pattern on this page to fit your foot. Then trace and cut out fifteen sandals from different colors of construction paper for each foot. Glue the fifteen sandal pieces carefully—one on top of the other. While still wet, form the top of the sandal to the shape of your foot. Allow to dry.

Sandpaper the edges of the sandal on an angle so that the rainbow of colors of paper will show.

Try on your sandals and wear them indoors for hours of fun!

Leather pieces could be substituted for the construction paper soles. Then only one piece of leather would be needed for each sandal.

Using colored sand from a florist, pet shop, or that you have colored yourself (directions from Sand Painting), pour layers of sand into a baby food jar. For added features, push a pencil or thin stick along the edge of the glass. This will let the sand slide down in a V.

Be sure to fill the jar **completely** and glue the lid on so it cannot be spilled.

When studying the American Desert, you could plant a little cactus plant in the jar. You may need a layer of charcoal for drainage. Consult a florist shop for their expert advice!

SAND TERRARIUMS

SAND PAINTING

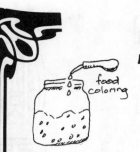

food coloring

Color sand using several drops of food coloring. Place the sand and the food coloring in a glass jar. Cover and shake to mix the color. Spread the damp sand onto a newspaper to dry. While this is drying, draw a simple desert scene. Spread glue onto the picture and sprinkle the sand on top. Fill in the entire picture with sand. When the sand painting has dried, gently tap the excess sand into a waste container.

If you have never worked with copper sheets, you will be amazed at the beautiful finished pictures everyone will be able to create!

Cut pieces of thin copper sheeting to about a nine-inch square. Place about one inch of newspaper under the copper square for padding.

On a nine-inch piece of construction paper, draw a **very detailed** picture of the sun. It could have a smile, frown, squinty eyes, nose, etc.

Place the drawn sun picture on top of the copper tooling. You may want to paper clip it in place so it will not move. Then using a pencil, Popsicle stick, or candy apple stick, press the outline of the sun onto the copper sheet. Continue to press the details into the picture. Remove the paper. Rub the large areas with flat sticks, such as Popsicle sticks, to give them sunken areas. Turn the copper over and rub on the back. This will give the front of the picture a raised effect.

A solution of liver of sulfide rubbed over the entire copper square will darken the picture. Gently rub off the dark areas, leaving some areas to accent your picture. Use fine steel wool to rub and polish the copper.

Smiling Suns

The Egyptians are happy people who enjoy laughing and having a good time! Holidays mean a great deal to the people. Family picnics in the countryside are a way of celebrating a spring festival known as **Shamm al-Nasim** (Smelling the Zephyr). It is a special holiday devoted to enjoying the soft wind!

Family holidays are important to everyone, too. Celebrations are held when babies are born, especially male babies. A wedding celebration may last for several days, and a funeral is accompanied by the crying of the women. The women cry in unison. This is known as **zagharett.**

If the family is of the **Muslim** religion, they observe the month-long festival of Ramadan. During this time the people are forbidden to eat or drink during the day. This is called fasting. When this month-long fasting is over, there is a new celebration with much feasting!

Festival of Fasting

Have you ever been forbidden to do something? What?_____

_____

How do you feel about this? _____

_____

How did you feel when it was over? _____

_____

Do you think the people **fast** only because they have to? Why?

_____

What could you give up if you were asked to? _____

_____

Fascinating Foods

Lamb is the most popular meat served in Egypt. Other dishes are made of dried beans or lentils. These dishes are baked for twenty-four hours and served with butter, olive oil or lemon. There is also mincemeat and green soup called **mulukhia.** Vegetables are plentiful and used abundantly in cooking. Fruits such as figs, dates and apricots, along with cheese, are usually eaten for dessert. Turkish coffee which is thick and sweet, along with tea, is an important drink.

Cooked foods are broiled or deep-fat fried. All cooking is done on a one-burner kerosene stove called a **wabur.** Among the poorer Egyptians, meat is rarely eaten. The main food is a large bean somewhat like a lima bean. Also, bread made from corn and wheat is eaten with tomatoes, lettuce, hard-boiled eggs and onions.

On the brass plate below, write a recipe for candy using dates, or draw the foods that would be part of a typical Egyptian meal.

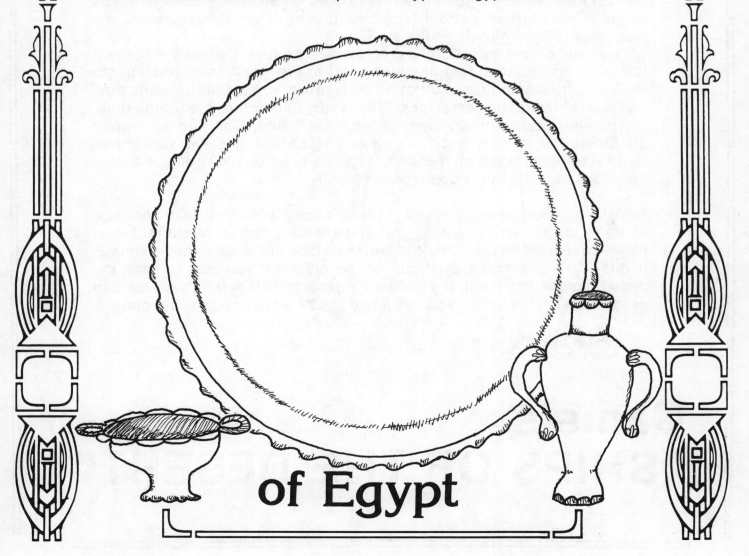

# of Egypt

The Arabian or one-humped camels are found on the Egyptian deserts. Some people call them **dromedaries.** They are more than seven feet tall at the hump and can weight up to 1500 pounds. Their short fur helps block out the hot sun. The most outstanding feature is the hump on the camel's back. The hump can weigh eighty pounds or more. Camels store fat in their humps. This fat is used as a food reserve when food is not available to the camels.

Camels can go for as long as three to four days without water. Even if water is available, camels will only replace the water they have used. A thirsty camel can drink up to thirty gallons of water in ten minutes. Camels do not lose water from their bodies through sweating.

A camel's eyes are protected from the sun and sandstorms by its long lashes. Its nostrils can be closed tightly so no sand will penetrate. The camel is not particular about food and its long tough lips and teeth will help keep it from getting indigestion.

The pads on the camel's chest and knees protect it when kneeling on the sand. Its two-toed feet are cushioned to prevent it from sinking into the sand. Its legs on one side of its body move forward at the same time and tend to make the camel rock. This is very uncomfortable for the rider.

A camel has an ornery disposition. It will bite, kick and spit when disturbed. But, as ugly and mean as a camel can be, it is also very important for transportation on a desert. It can carry up to 1000 pounds. It may not move quickly, but it does move steadily.

Make a desert caravan scene in damp sand placed in a cake pan. Add as many details as necessary using sticks or a pencil. Mix a recipe of plaster of Paris and pour over the entire surface of the sand (approximately one-half inch). Add a wire hook to the plaster if you plan to hang the desert picture. Wait until the plaster hardens and lift carefully off the bed of sand. Brush the extra sand away and you're ready to hang your picture.

Camels-
"SHIPS OF THE DESERT"

Color the costumes below.

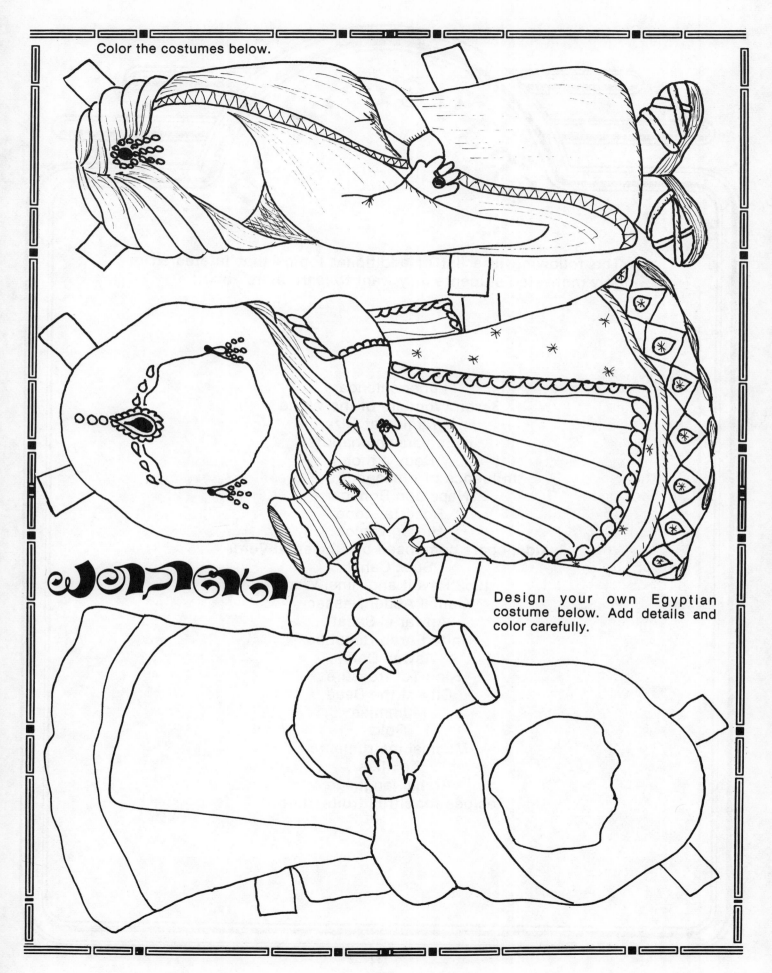

Design your own Egyptian costume below. Add details and color carefully.

111

travelin' on...

The following is a list of additional topics that interested or highly motivated students may want to learn more about:

Lunar calendar
Seven Wonders of the World
Pyramids at Giza
Horse-drawn chariots
Body armor
Influence of Alexander the Great
Napoleon Bonaparte
Rosetta stone
Hieroglyphics
Aida, opera composed by Giuseppe Verdi
Suez Canal
1952 Revolt and Sinai War
Gamal Abdel Nasser
Anwar el-Sadat
United Arab Republic
Aswan Dam
King Tut treasures
City of the Dead
Mummies
Gold
Musical instruments
Textiles
Arabic language
Recipes for citrus fruits, dates

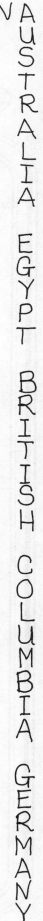

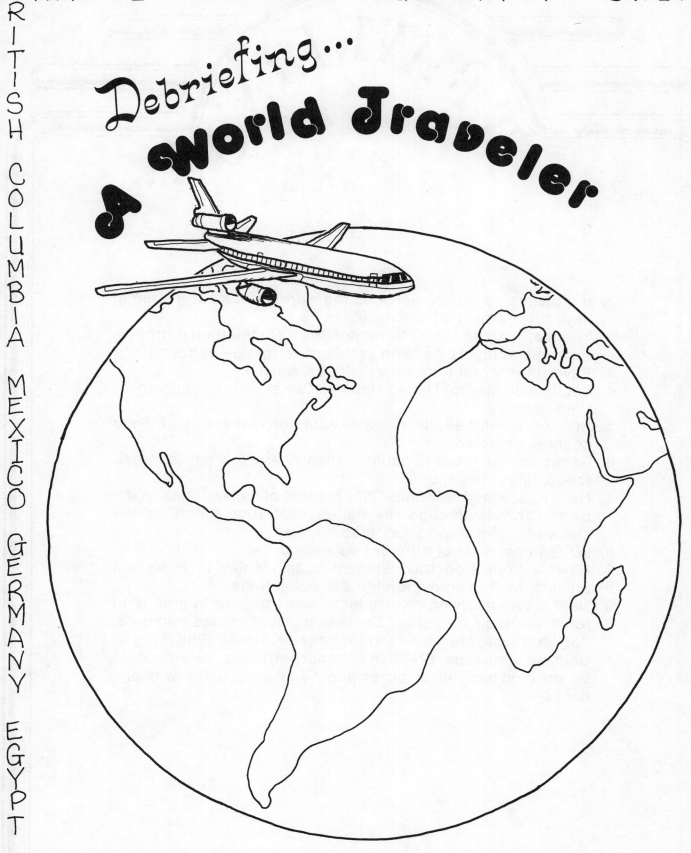

Debriefing...
A World Traveler

TRAVELIN' ON...

1. If students have accomplished the work for each of the units, have them attend a travelogue.
2. You may want to "sell" travel posters from the tourist offices. Have the students do math problems in foreign currency.
3. Review the world time zones with an activity.
4. Set up a display for United Nation's Day at your local library or travel agency.
5. Rent or obtain free films from the foreign embassies of these countries studied.
6. Set up booths for each country. Display artwork, written work, foods, flags, maps, etc.
7. Have a parade of countries. Play records of native songs as the parade travels through the hallways of your school or the sidewalks of your neighborhood.
8. Write to pen pals of different countries.
9. Have a Collection Day. Students can display their foreign stamps, foreign coins, foreign doll collections.
10. Hold a World Olympics. Students can compete in groups to represent their countries. Some suggested games might be: dog sled race, Mexican taco tack race, Chinese Ping-Pong or Chinese jump rope, Brazilian coconut carry, etc. Awards could be on gold and silver paper and hung around the winners' necks.

Tips for the Tour Guide

Klass Kulmination:

Plan a United Nations Day as a finale. Invite other classes, parents, and friends. Set up committees of children to work on separate parts of the activities. Several ideas that could be incorporated into your finale follow:

1. Have small groups design a float for their country to be used in a United Nations' parade. It could be built on a child's wagon. Include the flag and colors of the country.
2. Have a group enlarge a map of the country, showing major cities, physical features, products, etc. This could be used as a wall decoration.
3. Groups could make small flags attached to toothpicks. These could be given to guests as favors or prizes for games.
4. Groups could learn the games listed on the game page and teach them to guests. Other games could be used if you prefer.
5. Food could be made at school or brought from home. A tasting party of bite-sized samples is always a hit!
6. Groups could write a short skit telling about the special features of each country. A holiday is always a good theme.
7. Groups could sing the national anthems of other countries or child-written songs. You may want to include the U.S. national anthem.
8. A travelogue done by a local travel agency, or parent or friend who has traveled could be a special attraction.
9. If children make their own filmstrips or slides of different countries, a parent audience is always appreciative.
10. Learn the Disney song, "It's a Small World." Sing at the beginning of the program with the parade and at the end of the activities as the finale.

AN INVITATION

COVER

Come on a

World-Wide Tour

Fill in the information and use this invitation to invite your friends and parents to your Foreign Finale.

INSIDE

Date:

Time:

Place:

A Foreign Finale

FOREIGN FINALE

TRAVELOGUE

Have the children make short filmstrips using the filmstrip sample on the right. Have each child outline what he/she will draw and say for each frame. Include a title and frame with his/her name on it. This is an excellent sequencing activity.

If commercial film is available for this activity, be sure to use permanent markers. Typing will also work on the film. Add the narrative on a cassette tape.

Draw a picture or find a picture in a travel brochure the size of the slide rectangle. Write the dialogue to be used with the slide. The dialogue could be recorded on a cassette recorder, also.

Pictures for slides could also be drawn on 8″ x 11″ drawing paper. Then a slide camera could be used to photograph the drawings. Be sure to allow enough time to have the slides developed.

FOREIGN

China

FORTUNE COOKIES

4 egg whites
½ cup melted butter
¼ tsp. salt
2 tbsp. water

1 cup sugar
½ cup flour
½ tsp. vanilla
Small papers with fortunes written on them.

Directions: Mix sugar into the egg whites and blend until fluffy. Melt the butter and cool it so it's not too hot. Add the flour, salt, vanilla, water, and butter to the sugar mixture. Beat until smooth. Carefully grease cookie sheet very well. Using a teaspoon, pour batter into circles about 3 in. or 8 cm. in diameter. Bake at 375 degrees for 8 minutes. Place written fortune on each circle, fold it into thirds, and bend it gently in the center. Work quickly. Cookies may be reheated if they cool too quickly to get the correct shape. Makes about 30 cookies.

Mexico

SOPAIPILLAS

2 cups flour
1 tbsp. shortening
½ tsp. salt

3 tsp. baking powder
½ cup lukewarm water

Directions: Sift together dry ingredients. Cut in shortening until mixture resembles cornmeal. Gradually add water, stirring mixture with fork. Dough will be crumbly. Turn dough out onto lightly floured surface. Knead until smooth ball forms. Pull out 40 small balls from the large dough ball. Flatten with floured hands. Deep fry these flat circles 3 or 4 at a time for about 30 seconds on each side. Drain on paper toweling. Serve warm with honey and powdered sugar or roll them in cinnamon and sugar. Yields about 40 sopaipillas.

Brazil

CHOCOLATE-DIPPED BRAZIL NUTS

2 squares of bitter chocolate
¼ cup evaporated milk
1 tsp. vanilla extract

1 cup light corn syrup
½ cup granulated sugar
Brazil nuts, fresh fruits, or marshmallows

Directions: Melt the chocolate over a double boiler. Add corn syrup and evaporated milk to melted chocolate. Mix well. Remove from double boiler and place chocolate pan directly over low heat. Let chocolate boil slowly for 15 minutes. Stir occasionally. Add sugar and boil chocolate for 5 more minutes. Remove from heat. Cool a few minutes. Add vanilla and mix well. Refrigerate chocolate for at least 1 hour to thicken. Then cover a cookie sheet with waxed paper. Dip nuts into chocolate and place on paper. Refrigerate several hours to harden.

FOOD FAIR

British Columbia

BAKED CANADIAN BACON (for breakfast)

6 whole cloves
1 6-oz. pkg. sliced Canadian Bacon
½ tsp. dried mustard
½ cup apple juice

4 apples
½ lemon
½ cup maple syrup

Directions: Heat oven to 375 degrees. Lay bacon flat on the bottom of a casserole dish. Put cloves on top. Peel, core, and slice apples on top of the bacon. Squeeze lemon juice over apples. In another bowl, mix mustard with apple juice. Add syrup. Pour over bacon. Bake for 45 minutes.

Egypt

KAFTA (Meat Patties)

1 large onion
1 lb. ground beef
½ cup milk
½ tsp. salt
½ tsp. basil
1 cup bread crumbs

2 eggs
1 lb. lamb
¼ tsp. cayenne red pepper (measure carefully-it's hot!)
1 tbsp. chopped dry parsley
¼ tsp. marjoram
4 tbsp. butter

Directions: Peel onion, chop, set aside. Beat eggs in small bowl. Set aside. Mix all other ingredients together except the butter. Shape like hamburgers. Melt butter in frying pan and fry patties for about 10 minutes on each side.

Germany

PRETZEL INITIALS

1½ cups lukewarm water
4 cups flour
1 tsp. salt
1 egg
⅓ cup coarse salt

1 pkg. yeast
1 tbsp. sugar
2 tbsp. butter to grease cookie sheets
1 tbsp. water

Directions: In large mixing bowl add yeast to warm water. Let stand 5 min. until bubbly. In small bowl, combine 3 cups flour, sugar and salt. Stir until blended and dough forms a ball. Place dough on floured surface. Begin kneading, adding the last cup of flour. After 5 min. the dough should be smooth, not sticky. Tear into 20 small pieces. Roll each into a snake about ½" in diameter and 15" long. Shape into child's initial. Preheat oven to 425 degrees. Grease cookie sheets. Allow several inches between pretzels. In small bowl, combine egg and 1 tbsp. water. Mix well. Paint egg mixture on pretzel top and sprinkle with coarse salt. Bake at 425 degrees for 20 min. Cool on wire racks.

Games

Brazil

BAMBOO BOTTLE CAP BOUNCE
Materials: 12" - 18" dowel the thickness of a broomstick or bamboo pole
10 bottle caps
5 pennies
Directions: Draw a six-inch circle around the base of the dowel. Carefully place a bottle cap on the top of the dowel. Players stand five feet away from the dowel and try to knock the bottle cap off with a penny. If the bottle cap falls outside the small circle, one point is scored. If it falls inside the circle, or it is missed, it is the next player's turn. The player reaching ten (or twenty-five) points first is the winner!

Egypt

SILENCE IS GOLDEN
Materials: none
Directions: Divide into two teams. Form two circles. Choose a king (leader) for each team. The king lightly tickles the next player on the left and so on around the circle. When the action is all around the circle, the king starts a new action. This continues until someone on the team makes a sound. The team keeping silent the longest is the winning team.

Australia

LAST COUPLE COME
Materials: none
Directions: Each player finds a partner and lines up behind the extra player, who is "It." The couples should leave 6 to 10 feet of space between them and "It." To begin the game, "It" shouts, "Last Couple Come." The last couple sneaks up on "it." They can move quickly or slowly, one on each side of the line of players. "It" may not turn his head, but look only straight ahead. When they get even with him, he may begin to chase the couple. They must then try to grab each others hands to be safe. If a player is tagged, he and "It" become the head couple. If the couple grabs hands, they become the head couple and "It" must try again. This game is a favorite in many countries.

Germany

BASTE THE BEAR
Materials: 24" piece of rope, knotted at both ends
Directions: You may have ten to thirty players in a circle formation. The bear sits on a stool in the center of the circle. The keeper is with him. Both hold a knotted end of a piece of rope. Other players try to tag the bear without being tagged by the bear or the keeper. When a player is tagged, he takes the bear's place; the bear becomes the keeper and the keeper becomes a player. The bear and keeper may not let go of the rope during the game.

Games

China

CHINESE FINGER GUESSING
Materials: none
Directions: Two players face each other. They count "One, two, three." They put out their right hands, either closed or with one or more fingers extended. At the same time, they call out some number. The player who guesses the correct or closest number of the sum total of the extended fingers, scores a point. Five points may constitute a game. For a real challenge, have 3 or 4 players work together.

British Columbia

FOX AND GEESE
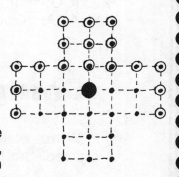

Materials: paper with this design enlarged on it
seventeen counters representing the geese
one larger counter representing the fox
Directions: The geese are placed on the circles marked with a circle and dot. The fox is in the center circle. One player moves the geese, the other, the fox. The fox starts first. He tries to move into position to jump the geese (like the checkers game). When a goose is jumped, it is removed from the board. The fox may jump forward, backward, up, down, and across. Players take turns and **must** move each time. The geese try to pen the fox so he cannot move. The geese may not jump; only the fox can.

Mexico

COYOTE AND SHEEP
Materials: none
Directions: Eight to twelve players are needed: one shepherd, one coyote, and the rest are sheep. The sheep form a line behind the shepherd, hands on the waist of the player in front of them. The shepherd is at the head of the line. When the coyote approaches, the shepherd says, "What does the coyote want?" The coyote answers, "I want fat meat!" The shepherd calls, "Then go to the end of the line; the fattest lambs are there!" The coyote then tries to get to the last sheep. The shepherd in turn must keep the coyote away. The sheep, still holding waists, must stay away from the coyote without breaking their line. If they do, the shepherd becomes the coyote, the first sheep becomes the shepherd and the same thing happens if the coyote tags the last sheep.

NATIONAL ANTHEMS

BRAZIL: "Hino Nacional"
WEST GERMANY: "Deutschland-Lied"
EGYPT: "El Salaam El Gomhoury"
AUSTRALIA: "God Save the Queen"

CHINA: "The March of the Volunteers"
MEXICO: "Himno Nacional de Mexico"
CANADA: "God Save the Queen" or "O Canada"

Compose Your Own Songs

Write songs of your own using familiar tunes. To help you get started, the first lines have been written for you.

EGYPT:
"Old MacDonald"
Old Ancient Egypt,
With secrets to tell...

AUSTRALIA:
"Twinkle, Twinkle
Little Star"
Kangaroos can jump up high...

MEXICO:
"Three Blind Mice"
Three centavos,
Three centavos,
See what they'll buy...

CANADA:
"She'll Be Coming 'Round the Mountain"
Grizzly's comin' down the Rockies when he comes...

GERMAN:
"Here We Go 'Round the Mulberry Bush"
Here we go into the Black Forest dark...

CHINA:
"Jack and Jill"
Chopsticks and rice
Are tasty and nice...

BRAZIL:
"The Old Gray Mare"
The deep dark jungle
Is full of ad-ven-ture...